ON THE TRACK OF UNKNOWN ANIMALS

ON THE TRACK OF UNKNOWN ANIMALS

by Bernard Heuvelmans

Translated and abridged by RICHARD GARNETT

With 78 drawings by MONIQUE WATTEAU

The MIT Press Cambridge, Massachusetts

To Dr. Serge Frechkop, who first led me into the muddy field of mammalogy, in deep gratitude and in the hope that he will not take amiss this excursion on the frontier of science and fantasy.

First MIT Press paperback printing, April 1972

Library of Congress Cataloging in Publication Data

Heuvelmans, Bernard.
On the track of unknown animals.

Translation of Sur la piste des bêtes ignorées.
1. Vertebrates. 2. Animal lore. I. Title.
[QL605.H413 1972] 596 72-521
ISBN 0-262-58020-9 (pbk.)

Note on This Edition

This abridged edition, intended for the general reader, is based on the revised edition of 1962 in which the author wrote,

> I have not been able to bring it completely up to date, for, while this would not have involved much rewriting, it would have meant adding so much unpublished material that has come to light in the meanwhile that the book would have been swollen to twice its size.

Since the present edition has had to be reduced from 558 pages to 320, to add anything new would have been absurd. Indeed, a certain amount of technical information has had to be omitted for lack of space. For instance the student will find in the complete edition a 25-page bibliography giving details of all the authorities referred to, and all the sources of quoted matter, as well as detailed acknowledgments to those who have helped in the writing of the book. This has all had to go by the board. I hope that what remains will prove to have been worth preserving in its new form.

R.G.
1965

CONTENTS

Page

PART FIVE

The Giants of the Far North

PART SIX

The Terrors of Africa

PART SEVEN

The Lesson of the Malagasy Ghosts

Drawings

by Monique Watteau

Maps

by K. C. Jordan

Photographs

(*Plates appear following page 146*)

PART ONE

THE GREAT DAYS OF ZOOLOGY ARE NOT DONE

And no one has a right to say that no water-babies exist, till they have seen no water-babies existing; which is quite a different thing, mind, from not seeing water-babies; and a thing which nobody ever did, or perhaps ever will do.

"But surely if there were water-babies, somebody would have caught one at least?"

Well. How do you know that somebody has not?

"But they would have put it into spirits, or into the *Illustrated News,* or perhaps cut it into two halves, poor dear little thing, and sent one to Professor Owen, and one to Professor Huxley, to see what they would each say about it."

Ah, my dear little man! that does not follow at all, as you will see before the end of the story.

CHARLES KINGSLEY, *The Water-Babies.*

CHAPTER 1

THERE ARE LOST WORLDS EVERYWHERE

Even to-day there are in every part of the world inaccessible areas where there certainly exist wild creatures unknown to naturalists.

JOSEPH DELMONT, the great animal-catcher.

MOST zoologists are skeptical about the possibilities of discovering new species of large animals, and some of them do not, with legitimate scientific skepticism, keep an open mind until the species is proved to exist, but categorically deny that it can possibly do so until they have been forcibly proved wrong. Their obstinacy is based on three propositions: the world has now been completely explored; no new animals have been discovered for a long time—at least not since the okapi; and many of the animals alleged to exist are fossil species and therefore long extinct. All three propositions are fallacies, as I hope to show in the first three chapters of this book.

The world is by no means thoroughly explored. It is true that there are no more large islands or continents to be discovered. But because a country is on the map it does not mean that we know all about its inhabitants.

Let us begin with the most unknown but least hopeful continent, Antarctica. It covers 5,000,000 square miles, the area of the United States and western Europe put together, and I doubt if more than half of it has been explored. Of course it is the most inhospitable country in the world, and boasts no known terrestrial mammals. The only amphibious ones, the seals, rarely venture far from the coast, nor do the penguins, petrels, gulls, and skuas, which are all sea birds. There is little hope of finding unknown mammals or birds in the middle of Antarctica, but there is a faint chance.

For instance *Younyi Technik* of April 1958 reported that a Soviet expedition from Mirny found a warm lake at an altitude of 130 feet, 8 miles inland on the Antarctic continent in which there were baby seals less than a month old. Presumably they had been born there, but even so it is almost impossible to imagine their mothers crawling 8 miles—and what could they find to feed on? No doubt Antarctica holds other surprises.

Greenland is hardly any better known. But all this country has frequently been flown over by low-flying aircraft, from which you can clearly see seals lying in the sun and polar bears which are sometimes found 500 miles inland. On this white background everything stands out like a flea on a white sheet. This is why Paul-Émile Victor, the great French polar explorer, writes: "Perhaps one day hitherto unknown species of animals will be discovered in these regions. But . . . they will only be small animals, like insects for instance."

Africa is infinitely better explored, but holds more promise for the naturalist. Although none of it is utterly unknown, except for a few patches in the middle of the Kalahari Desert, much of it has been mapped but not explored. The use of aerial photography gives the illusion that the world is much better known than it is. For there is not the least hope of finding out from an aerial photograph of forests or steppes of tall grass and bush what animals live there, even if they are as large as an elephant or a rhinoceros. There is even less if they are aquatic animals like the hippopotamus, the manatee, and most of the large reptiles.

The likeliest country to contain large unknown animals is the great equatorial rain forest in which six of the most important zoological discoveries of the present century have been made. The American naturalist Herbert Lang writes of it:

> The numerous sportsmen who had visited nearly all parts in Africa found no attraction in these forests. . . . The immensity of the wilderness is appalling; for over eighteen hundred miles without a break it stretches more than half-way across the continent, from the coast of Guinea to the Ruwenzori. In spite of tropical luxuriance, it is one of the most dismal spots on the face of the globe.

Other little-known areas in tropical Africa are the mountainous parts of Kenya and Katanga and the areas of swamp or scattered lakes which might shelter large unknown animals; that is to say, the Rhodesian lakes, the Addar marshes, which cover 1,500 square miles, or those of the Bahr-el-Ghazal, a tributary of the

White Nile in Sudan, which are even more vast and some parts of which have never been crossed.

There may be much for the naturalist in the foothills of the Himalayas and the valleys in the north and southeast of this great mountain range and in the vast Siberian coniferous forest, the taiga. In 1955 Soviet geographers were surprised to discover that whole ranges of mountains in the Kolyma and Indigirka areas, the general shape of which was supposed to be known, actually stretched in quite unexpected directions. Where mountains can so long remain unknown, there is good hope for the naturalist.

Zoologists have far from exhausted the wealth of the tropical jungle in such supposedly known lands as India, Burma, the Moi country in Indochina, Malaya, central Borneo, and Sumatra. Nothing whatever is known about the mountainous center of New Guinea except that various tribes, many of them Pygmies, still live there in the Stone Age. In 1938 the American naturalist Richard Archbold accidentally discovered in the western part of the island an excellently irrigated valley inhabited by 60,000 people. And in June 1954 patrol aircraft found quite unknown tribes in valleys in the southwest. Their population was estimated at 100,000, a third of the number of Papuans already known.

Most of the interior of Australia is covered with deserts of sand, salt, and thorny bush. Even the least inhospitable parts are steppes of tall grass scattered with stunted shrubs. Hardly anybody goes there except a few prospectors, who come back with tales of animals so fantastic that they are usually thought to be drunken visions. But anything can happen in a continent where there are ranges of mountains which have never been seen except from an airplane.

The continent that holds the most mysteries is certainly South America. There are dozens of blank spots on the map. On the more accurate surveys the Amazon basin is as full of holes as a Gruyère cheese. Bertrand Flornoy, who has spent much of his life exploring this country, calls it "a vast night of trees which stretches for two million square miles." Four centuries of exploration have not got to the bottom of it, because, as Flornoy says:

> The virgin forest constantly opposes man's efforts with its irresistible power which transforms all decay into new life in a matter of hours. A path is no sooner opened than the jungle wipes it out. And, if you fly over the Amazonian forest, towns like Belem and Manaos in Brazil and Iquitos in Peru look like clearings. As to the clearings in which the Indians live, they are invisible.

Besides the heart of the Matto Grosso, a large part of Colombia is unknown, at least in practice, and so is the hinterland of Venezuela and Guiana and a large chunk of the Cordillera of the Andes. The wooded part of Patagonia is far from being thoroughly explored.

There is one part of South America which is of special interest, since it is a perfect example of an almost inaccessible nature reserve. It lies in that corner of Venezuela which includes the sources of the Orinoco. There, among the almost impenetrable jungle of the Gran Sabana, rise the mesas, vast limestone plateaus cut off from the rest of the world by sheer cliffs between 3,000 and 10,000 feet high. Some of these mesas are almost 20 miles long, great islands of thick vegetation about which often nothing is known. On those which have been explored rodents have been found that are unknown in the local forests but recall those of the Venezuelan Andes hundreds of miles to the west. In any case these plateaus are not easy to reach. In 1937 Jimmy Angel tried to land his aircraft on the top of the Auyan Tepui mesa and on this same mesa more recently discovered a waterfall fifteen times higher than Niagara. From the sheer 5,000-foot wall of Auyan Tepui gigantic columns of water fall 3,200 feet. As the top of the waterfall was usually hidden in clouds, the Taurepan Indians long ago told the conquistadors that there were terrible waterfalls that fell directly from the sky. Naturally they were not believed.

If the highest waterfall in the world should have remained so long unnoticed, how can we be sure that this country does not still conceal some unknown animals, even quite large ones? I do not suggest that we can hope to emulate Conan Doyle's *The Lost World,* in which Professor Challenger found on a similar plateau *all* the enormous reptiles from all parts of the world in the Mesozoic era and even some ape men. But it is absurd to insist that there can be *no* unknown animals in such little-known country.

In any case, animals of a past age do not need to be imprisoned on a small island or on the top of a mesa to be preserved from destruction by their subsequent competitors. A large island will do as well. Australia is the marsupials' "lost world"; Madagascar the lemurs'; and New Zealand that of the rhynchocephalous reptiles and a once-flourishing empire of birds. Even on continents a particular habitat, unsuitable for most other creatures, can provide a very safe retreat for the species that have managed to adapt themselves to it.

But in fact for the zoologist the whole world is terra incognita, and there is still hope of finding many new species—especially of small animals.

The entomologist has the largest haul, which is hardly surprising, since three quarters of a million kinds of known animals belong to the troublesome insect empire. On the average 5,000 new species or subspecies of insects are described each year. Discoveries of new land vertebrates—which are what we are concerned with here—are obviously less frequent, since there are altogether only 60,000 species and subspecies of amphibians, reptiles, mammals, and birds known today. The following table will give some idea of how the numbers of known forms have increased since Linnaeus made his fundamental classification. The subsequent stages are marked by the works of those who have attempted a census or estimate of this kind. The figures for 1960 are approximate and have been obtained by extrapolation, producing the curve formed by the earlier figures.

	Linnaeus (1758)	Cuvier (1817)	Leunis and Ludwig (1886)	Karl Möbius (1898)	W. Arndt (1939)	? (1960)
Amphibians & Reptiles	181	239	3,400	5,000	5,461	6,000
Mammals	183	386	2,300	3,500	13,000	17,000
Birds	444	765	10,150	13,000	28,000	37,000

Thus, since the beginning of this century an average of about 15 reptiles or amphibians, 220 mammals, and 400 birds have been described each year. Not all were really new and hitherto unknown, for many zoologists create new species and subspecies on the strength of barely perceptible differences. All the same, we can reckon that one in ten of these descriptions refers to a clearly distinct species. Therefore each year the catalogue includes some 40 birds and 20 mammals that were hitherto unknown. The former are usually small songbirds, the latter little rodents or bats or even small marsupials or insectivores. Reptiles and amphibians are much rarer and almost an exception: there are no more than one or two a year.

As soon as you set about looking for large animals the area of the world to be searched becomes much smaller, and is generally restricted to the places I have detailed above. All the same, even without Antarctica, the regions that are little known amount to no less than *one tenth* of the land surface of the globe.

CHAPTER 2

CUVIER'S RASH DICTUM

> You must not say that this cannot be, or that that is contrary to nature. You do not know what Nature is, or what she can do; and nobody knows, not even Sir Roderick Murchison, or Professor Owen, or Professor Sedgwick, or Professor Huxley, or Mr. Darwin, or Professor Faraday, or Mr. Grove . . . They are very wise men; and you must listen respectfully to all they say: but even if they should say, which I am sure they never would, "That cannot exist. That is contrary to nature," you must wait a little and see; for perhaps even they may be wrong.
>
> CHARLES KINGSLEY, *The Water-Babies.*

MORE than a century ago, in 1812, Baron Georges Cuvier was rash enough to pronounce that "there is little hope of discovering new species of large quadrupeds." The living fauna of the world was well enough known for him. The animals even of such newly explored lands as America and Australia had been numbered and described. Naturalists should now concentrate on extinct animals, remains of which were being disinterred in such quantities. Zoology was out of fashion, paleontology, which Cuvier had fathered, was all the rage.

This pronouncement, like others of his, would not have mattered had he not been such a despot. Brilliant anatomist though he was, the ill effects of his tyranny are still felt, and reactionary scientists still repeat his dictum, though it has been disproved over and over again.

In 1819 Cuvier had not long uttered his words when his pupil Diard wrote to him:

> When I first saw the tapir—of which I send you a picture—at Barrackpore, I was astonished that such a large animal should still be

unknown, especially as I had seen at the Asiatic Society the head of a similar animal which Governor Farquhar had sent there on 2 April 1806, saying that the tapir was as common in the Indian forests as the elephant and the rhinoceros.

Actually the white-backed tapir had been known to the Chinese and Japanese since time immemorial, but Cuvier always scorned the beliefs of simple people and even distrusted all travelers' tales. For him the tapir was a specifically American animal, although Wahlfeldt had mentioned a little black-and-white "rhinoceros" that looked very much like a tapir in Sumatra in 1772. Marsden had described it less equivocally in the same island in 1783. Sir Stamford Raffles had learned about it in 1805, and Farquhar had shot one near Malacca and described it to the Asiatic Society in 1816.

When Cuvier received Diard's letter and drawing he hastened to publish an official description, but he was too late. Someone else had beaten him to it.

This was only the first of a long series of large animals which were gradually discovered. I cannot hope to list them all here, but will confine myself to animals which were quite new or could not easily remain undetected.

Three primates deservedly open the list. In 1820 Desmarest announced the existence of the black or Celebesean ape (*Cynopithecus niger*), which is really a sort of would-be baboon with a very prominent snout. A year later Raffles described the largest of all the gibbons, the siamang of Sumatra, which required a genus of its own (*Symphalangus syndactylus*); it stands more than 3 feet high on its hind legs. And in 1835 Rüppel discovered the largest of the baboons, the gelada (*Theropithecus gelada*), in the rocky mountains of southern Abyssinia.

The water chevrotain (*Hyemoschus aquaticus*) described by Ogilby in 1840 from a specimen brought back from Sierra Leone is admittedly not a large animal—it is only a foot high at the withers—but this little ungulate is important because it is the only living representative of a large group which lived in Europe during the second half of the Tertiary era. It was the first of many living fossils to be found in the great equatorial rain forest.

In the same year an affair began which was not finally settled for another fifty years. The vice-president of Philadelphia Academy of Natural Sciences heard from a traveler who had returned from a long journey in Liberia, that in the inland rivers of that country there was a little hippopotamus hardly as long as a medium-sized heifer. The natives hunted it for its meat, and the traveler had not only seen the beast several times but had eaten it himself. Dr. Morton was rather skeptical about this story, but in 1843 his friend Dr. Goheen sent him a set of mammals' skulls from Monrovia, among which he was astonished to find two not unlike a hippopotamus's but very much smaller. He described them in 1849, naming the species *Hippopotamus liberiensis,* but his colleague Joseph Leidy soon showed that it differed from the ordinary hippopotamus in so many respects that it deserved to be put in a new genus, *Choeropsis.*

Despite the two skulls most scientists still denied its existence. Blainville wrote:

> It seems however, from what I hear from R. Owen who has seen the skull that the specific differences depend upon the size, and that the rest are of little importance.

Actually the pygmy's skull is only a little over half the size of the ordinary hippopotamus's and Leidy, who had studied it thoroughly, wrote:

> In the anatomical details of the upper part of the skull, the differences are so great between Chœropsis and Hippopotamus, that any

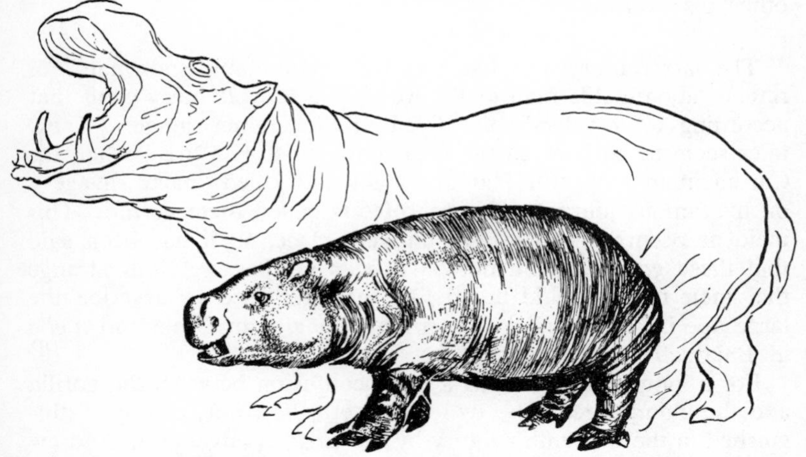

1. Pygmy hippopotamus, compared with the large species.

one finding a fragment constituted by this part of the former, from his previous knowledge of the latter alone, would not suspect it even of being closely allied to the genus Hippopotamus.

But "R. Owen" had said otherwise. And he was Professor Sir Richard Owen, the greatest British paleontologist, a fervent disciple of Cuvier and as brilliant an anatomist as his master, but none the less intolerant, prejudiced, and skeptical. And, unfortunately, like Cuvier's, his word was taken as gospel.

Around 1870 a young pygmy hippopotamus, which weighed barely 30 pounds, was sent to the Dublin Zoo, where it lived for several weeks. Twenty years later Johannes Büttikofer, curator of the Rijksmuseum at Leiden, came back from two long visits to Liberia with much information about the pygmy hippopotamus's habits and even several skulls and skeletons in a rather poor state.

At the beginning of the twentieth century there were at least a score of specimens—many very fragmentary, it is true—in museums in various parts of the world. Yet many naturalists still thought *Choeropsis* was just a young hippopotamus or an individual freak. There was even one natural history museum in which a badly mounted specimen was classified with the fossils—an original way of denying that the creature was extant!

Eventually, of course, its existence was established so that no one could deny it, but not for many years. For the moment let us leave the poor beast temporarily buried alive, and proceed to other discoveries.

The last century also had its own "abominable snowman," or rather "abominable man of the woods," a terrible hairy giant that according to travelers' tales lived in the Guinea jungles. These tales seem to go back to the fifth century before Christ, when the Carthaginian navigator Hanno fought and killed "hairy savages" on his famous journey. The interpreters called them *gorillas*. This incident occurred in a rather unlikely place, so it has been said that these *gorillas* could not have been gorillas, yet it is strange that their name should be so like that still used to describe the large ape in some dialects, for instance *n'giya* in Eveia and *n'gila* in Bakoueli.

For a long time there was a great confusion between the gorilla and the chimpanzee. The two apes were, however, clearly distinguished in the sixteenth century by the English adventurer Andrew Battel, who was taken prisoner by the Portuguese and spent many years in Africa, chiefly in the forests around the River Banya and Mayombe. His story was published in 1625 in a collection of travels entitled *Purchas his Pilgrimes.*

> The greatest of these two Monsters is called, *Pongo,* in their Language: and the lesser is called, *Engeco.* This *Pongo* is in all proportion like a man, but that he is more like a Giant in a stature, then a man: for he is very tall, a hath a mans face, hollow-eyed, with long haire vpon his browes. His face and eares are without haire, and his hands also. His bodie is full of haire, but not very thicke, and it is of a dunnish colour. He differeth not from a man, but in his legs, for they haue no calfe. Hee goeth alwaies vpon his legs, and carrieth his hands clasped on the nape of his necke, when he goeth vpon the ground. . . . They goe many together, and kill many *Negroes* that trauaile in the Woods. Many times they fall vpon the Elephants, which come to feed where they be, and so beate them with their clubbed fists, and pieces of wood, that they will runne roaring away from them. Those *Pongoes* are neuer taken aliue, because they are so strong, that ten men cannot hold one of them. . . .

Purchas adds that Battel, who was his neighbor,

> told me in conference with him, that one of these *Pongos* tooke a *Negro* Boy of his, which liued a moneth with them. For they hurt not those which they surprise at vnawares, except they look on them, which hee auoyded. He said, their highth was like a mans, but their bignesse twice as great. I saw the *Negro* boy.

It was not until the middle of the nineteenth century that the American Protestant missionaries Savage and Wilson collected the usual legends about the large ape in Gabon, and also some skulls which found their way back to America. In 1847 Savage in collaboration with Professor Jeffries Wyman of Boston published a first brief description of the gorilla. Today, after a century of research, the "abominable man of the woods" has not entirely lost the largely unjustified aura of terror with which it was surrounded in the legends.

When B. H. Hodgson was exploring Tibet and the neighboring countries in 1850 he was given three skins and several skulls of a sheep with gray wool which seemed to be as large and heavy as a small buffalo. He christened it *Budorcas taxicolor.* It was well known to the Mishmis of Assam as *takin,* but Hodgson was never able to see one himself. We now know three quite different species of takin. The golden takin (*B. bedfordi*), not discovered until 1911, is the most interesting. It is very rarely seen, for the good reason that it lives in eastern China between 8,000 and 14,000 feet, and spends most of the daylight hours in thickets of rhododendron and dwarf bamboo, emerging to browse on the grassy slopes only at dusk.

The first takin brought back alive from Assam to Europe was a male which was presented to the London Zoo by J. C. White on June 22, 1909, 59 years after the beast was discovered.

Schomburgk's deer (*Rucervus schomburgki*), which was described by Blyth in 1863, is remarkable not only for its curiously forked antlers which may boast as many as 20 points, but also because it has never been seen in its natural habitat in Siam by a single white man. All that is known about it has been learned from rare specimens captured by the Siamese or born in captivity. There is indeed little hope that any European will ever see it in its natural state, for it is quite likely that it has been extinct for several years.

Occasionally an intelligent naturalist with his ears open has been able to save a species. When the French missionary Father Armand David crossed eastern Asia between 1865 and 1869, he discovered three new kinds of large mammals.

When he was in Peking in 1865 he heard that sacred animals were kept in the Nan Hai-tzu Imperial Park several li south of the city. His curiosity was aroused, but he knew that entry to the park was strictly forbidden; so he eluded the Tartar guard by climbing the 45-mile wall that encircled the park and protected it from curious eyes. He was most excited by what he saw. Besides the animals that he knew well, there was a large herd—some 120 head—of very strange deer.

By bribing some of the keepers, Father David managed to procure a pair of antlers. The following year he obtained two skins in a fairly good condition. Then diplomats of several countries set about trying to obtain specimens, and thanks to the Imperial Minister Hen-Chi, the first three living specimens were presented to France. They did not survive the voyage, but Alphonse Milne-Edwards studied their remains and in 1866 christened the species *Elaphurus davidianus*. It is a graceful and sad-eyed deer, with light tawny-brown fur in a shaggy coat which rises to a mane on the neck. Unlike most other species it sheds its antlers twice a year.

Further specimens were sent to other countries and fared better than their predecessors. In 1895 floods destroyed a large piece of the wall around the park, so that some of the deer escaped and were eaten by the starving population.

2. Père David's deer.

The Boxer Rebellion in 1900 was even more fatal. The rebels invaded the Emperor's grounds and slaughtered all the deer but one—the only survivor wild in the whole of China, for the species had been completely exterminated several centuries before.

Fortunately Father David had climbed the wall, and thanks to him a few rare individuals of "Père David's deer" had been preserved in Europe, where the Duke of Bedford was able to save the species from extinction. In his park at Woburn Abbey, 15 breeding survivors have so multiplied that now, after half a century, there is a herd of 250, from which several zoos have been provided with specimens.

On Chinese vases and silk paintings you can sometimes see a sort of grinning demon with a turquoise-blue face and tail. All the front of its body is as red as fire, but it has an incongruous little snub nose. This beast was Father David's second great discovery. It turned out to be a not so very stylized picture of a monkey that lived in the snows of eastern Tibet, and this only increased the skeptics' incredulity, since everyone knows that monkeys prefer a warm climate. But this "snow monkey" (see Fig. 26) was found at altitudes of over 10,000 feet. In 1870 Milne-Edwards, after examining its skin, christened it *Rhinopithecus roxellanae* after Roxellana, the beautiful snub-nosed slave who married Suleiman II and thus showed that fairy tales could sometimes come true.

Father David was taking tea and eating sweets with a rich landowner in Szechwan on March 11, 1869, when he made his third great discovery. In his host's house there was a curiously marked fur which he recognized as that of "the famous black-and-white bear" which he had several times heard tell of. As far as I know, the first mention of it is in a manuscript dating from A.D. 621, during the reign of the first of the Tang emperors.

Father David was convinced that this legendary animal existed, and urged the Chinese hunters to get him a specimen. Twelve days later, on March 23, they brought him a young one which they had caught alive, but had killed in order to carry it home. Thus he was the first Westerner to obtain the fur of what we now call the giant panda. It is white all over its body except on its legs, the top of its chest, its shoulders, its ears, the tip of its nose, and around its eyes, where it is dark brown, almost black. Unlike a true bear, it has hair even on the soles of its feet.

On April 1 the hunters brought Father David the skin and skeleton of an adult, and six days later he obtained a pretty little flame-colored animal already known to science, the panda (*Ailurus fulgens*).

Father David certainly never guessed that this kittenish little creature was related to the big black-and-white "bear" whose remains he had just collected, and it remained the "bamboo-bear" or "Père David's bear" for some time. But when Milne-Edwards had studied its dentition and its skeleton he realized with a touch of genius that it was related to the Procyonidae, the raccoons, and gave it the name of *Ailuropoda melanoleucus,* or "the black-and-white animal with panda's feet."

It only remained to find the giant panda in its natural habitat and capture a specimen. But this was sooner said than done. The giant panda had hardly been discovered when it seemed to disappear without leaving a trace. No more was heard of it for over half a century.

The crested rat (*Lophiomys imhausi*) is certainly the most conspicuous of rodents. It may reach a length of 16 inches and its body is covered with a coat of long black and white hairs, which stand up to make a striking crest along its spine. Yet this creature, which lives among the rocks in the mountains of East Africa, was not discovered until 1867, when it was immediately described by Alphonse Milne-Edwards, who was then having a remarkable run of luck. This rodent was so unusual that a whole new family had to be created to accommodate it.

Six years later another no less extraordinary rodent came to light and also added another family to this order. It was the *Dinomys branickii,* or pacarana, which is the size of a fox terrier. Yet it was not known until 1873, and then not particularly well. Professor Walter Henricks Hodge writes:

> For many years the Pacarana was known only from the original specimen that was captured in 1873 as it wandered about in a *hacienda* building at the hamlet of Vitoc, Peru, at the headwaters of the Perené River just east across the Andes from Lima. Until 1904 the species remained out of scientific headlines, but in that year two living Pacaranas were seen in Pará from the upper Rio Purus in Brazil. In 1922 a series of specimens was obtained from the forested Andean area to the east of Lima; and, finally, in 1925 the New York Zoological Garden obtained a living specimen for exhibition. In 1930 two additional Pacaranas were obtained and about the same time the species made its zoo debut in Hamburg and London. Thus all our present knowledge of this animal is based on less than fifty individual specimens. The information concerning this giant rat has remained in fact so scanty that as recently as 1942 it was included among a list of extinct and vanishing mammals of South America. Rare it seems to be but whether on the road to extinction is another question. Pacarana territory includes some of the most precipitous and consequently inaccessible and unknown forested country in the world. Only a few trails or roads cut into this terrain and only after this *terra incognita* is better known to field zoologists can we state with certainty the status of this rare rodent, which may be much more abundant than the few existing records would indicate.

Between 1869 and 1882 a good half-dozen new animals were added to the Ungulates, the order which includes most of the larger land animals. In 1869 Blyth described a new species of African antelope with corkscrew horns, the lesser kudu (*Strepsiceros imberbis*). In 1870 Swinhoe announced a new kind of deer in the Chinese marshes, the Chinese water deer (*Hydropotes inermis*), which was quite without horns. The following year Milne-Edwards kept up his brilliant series of descriptions with another Chinese deer, the tufted deer (*Elaphodus cephalophus*), which had mere stumps of antlers, but the male was armed with protruding canine teeth.

In 1872 and 1878 Sir Victor Brooke christened two African gazelles. The first, Grant's gazelle (*Gazella granti*), was merely a new species, but the second, Waller's gazelle (*Litocranius walleri*), belonged to a quite new genus. It is a most unusual-looking beast with a neck as long as its body—a sort of giraffe-gazelle. The natives call it gerenuk and have known it since time immemorial, and it appears in Egyptian bas-reliefs dating from the sixth century B.C.

All too often zoologists make the mistake of underestimating ancient or primitive art as a source of information. Had they studied Egyptian murals more carefully they would not only have noticed the gerenuk, but also a zebra with particularly narrow and numerous stripes. The Scottish explorer James Augustus Grant—after whom the gazelle was named—claimed to have seen one of these zebras in Abyssinia in 1860, but the experts were still skeptical.

In 1882, however, Menelik I, Emperor of Abyssinia, gave the French President, Jules Grévy, a zebra which exactly fitted Grant's description. When it was brought to France and deposited in the Jardin des Plantes everyone could see that it was a new species, which was larger than the other zebras and had a different pattern on its coat. The poor brute was worn out by its long journey and had a fit and died several days after its arrival. All the same, Oustalet was able to add another new large quadruped (*Dolichohippus grevyi*) to the world's fauna.

At about the same time the news reached Europe that the Russian explorer Przewalski had discovered a species of wild horse in western Mongolia. Actually, though he was a cavalry officer, Przewalski had taken it for an ass, for it had a large head and relatively short legs. I. S. Poliakoff, who published the first description of it in 1881, needed all his patience to convince him that it was really a horse (*Equus przewalskii*). It is the only living example of a truly wild horse—and not a descendant of tame

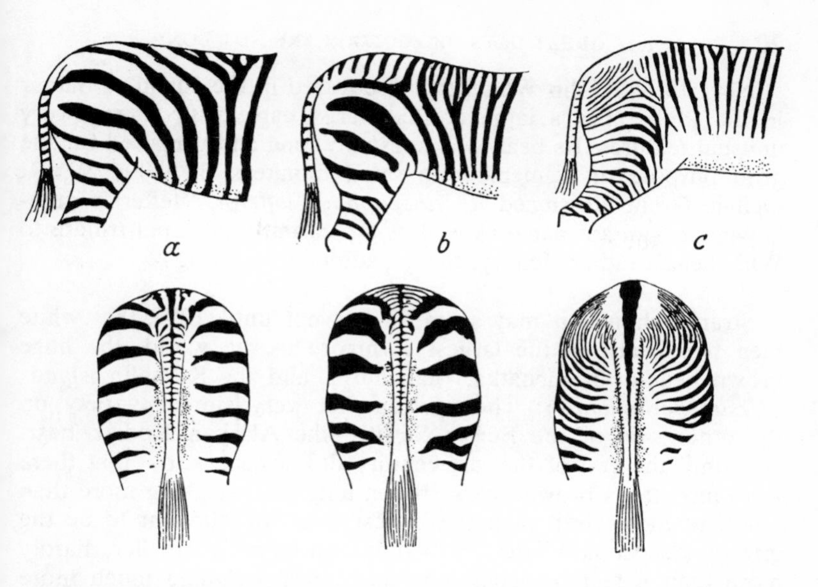

3. The peculiar pattern of the skin of Grévy's zebra (*c*), compared with that of Burchell's zebra (*a*) and the mountain zebra (*b*), (after Frechkop, 1947).

horses that have gone wild, like those in America. The very last specimen of the only other surviving wild horse, the tarpan (*Equus gmelini*) of the Polish forests and southern Russian steppes, died in the Ukraine at Christmas 1879. Hence the special importance of Colonel Przewalski's discovery.

In 1891 Oldfield Thomas christened another large quadruped, the dibatag antelope (*Ammodorcas clarkei*), commonly called Clarke's gazelle because it was discovered in Somaliland by W. H. Clarke.

Here I wish to add a large bird to the list. You may object that no bird, however big, can be a large quadruped, and that Cuvier never suggested that there was little hope of finding new species of large birds. True, but this was certainly because he thought there was *no hope at all*. There are no more conspicuous creatures than birds. Their ability to fly away at the first alarm means that they do not have to hide so much as mammals. For the same reason they can wear brilliantly colored plumage without suffering for it. Certainly it is very rarely that a large and quite unknown bird is discovered.

Nevertheless John Whitehead discovered in the island of Samar in the Philippines a large crested harpy eagle with several very unusual features. Its beak was unusually long and was used for the grim purpose of dismembering small monkeys. In 1896 W. R. Ogilvie Grant christened it *Pithecophaga jefferyi,* "Jeffery's monkey-eater," an apt name as well as a pleasantly informal tribute to Whitehead's father, Jeffery, the expedition's sponsor.

Strange though it may seem, it was not until 1898 that white men first heard of the largest carnivore in the world, the huge brown bear of Kamchatka, Manchuria, and the Sakhalin islands (*Ursus beringianus*). There is another very similar variety on the other side of the Bering Straits—the Alaskan Kodiak bear. Not until the end of the last century did science know that there were monstrous brown bears 10 feet long and weighing more than 1,600 pounds. Until then the largest bear was thought to be the grizzly (*Ursus horribilis*), which is considerably smaller, hardly more than 6 feet 6 inches long and never weighing much more than 1,100 pounds.

It was likewise not until 1900 that a variety of the second largest land animal was found to exist in part of the world where it had been quite unknown. The beast was the white rhinoceros * (*Ceratotherium simum*), which had been thought to be confined to South Africa south of the Zambezi. It is 15 feet long, more than 6 feet high, and may weigh more than 2 tons. Its longer horn may be 5 feet 2 inches high, the height of a small man. It is the most impressive of all the rhinoceroses.

Then Captain A. St. H. Gibbons brought a white rhinoceros skull from Lado on the Upper Nile, some 2,000 miles farther north than the area in which it had hitherto been found. Other specimens were collected by Major Powell-Cotton, enabling the great Lydekker to describe it as a new geographical race or subspecies of the white rhinoceros (*Ceratotherium simum cottoni*). Its area of distribution stretches from northeast of the Uele into the Belgian Congo. That such a huge beast could have remained unnoticed in a country generally thought to be well explored seemed almost like a joke.

* Actually the white rhinoceros is the same grayish hue as the black rhinoceros. The name arises from a mistranslation of the Boer name for the beast, which is *weid* rhino. This means "wide"—not "white"—and refers to its broad, square nose; for the essential difference between the two species is in the shape of the upper lip and the distance between the nostrils.

The next discovery was of the largest known ape, the mountain gorilla. Its existence was established in 1901, and two years later Matschie christened it *Gorilla beringei* in honor of Captain Oscar von Beringe, who brought the first skin back from Kivu on the eastern side of the Belgian Congo. Until then only one species of gorilla was known (*Gorilla gorilla*), whose habitat stretched from Gabon to the Cameroons and the French Congo. Yet as early as 1860 the natives of Rumanika in Ruanda (adjacent to Kivu) had told Speke of a monster which hugged women so savagely that they died, and the Negroes in the east of the Belgian Congo had

4. Mountain gorilla.

always been full of tales about it. But the whites refused to believe these "absurd legends" until Beringe killed the first specimen.

The Kivu gorilla is certainly a fearsome beast. It may be as much as 6 feet 6 inches high—taller than the coastal gorilla, which has never been known to exceed 5 feet 11 inches. An old male with its arms outstretched can sometimes span more than 9 feet, its chest measurement being 67 inches and its biceps 25 inches. Its weight may be as much as 600 and possibly 700 pounds. That such a huge brute could remain unknown until the beginning of the twentieth century shows what a shy and pacific beast it really is.

But the greatest sensation of the 1900s was the okapi. As Dr. Maurice Burton has said: one can have no idea today of "the romance surrounding the discovery of the Okapi, nor of the excitement caused in natural history circles, first by the vague reports of its presence, and later by its actual finding."

The first evidence of the okapi's existence was extremely slender, and consisted of no more than three lines in H. M. Stanley's linguistic notes about the Pygmies in *In Darkest Africa,* published in 1860.

> The Wambutti knew a donkey and called it "atti." They say that they sometimes catch them in pits. What they can find to eat is a wonder. They eat leaves.

No one had ever heard of wild asses in the Belgian Congo. The only Equidae known there were zebras, but these specifically running beasts never lived in the thick jungle inhabited by the shy Pygmies. So the experts were skeptical.

All the same, Sir Harry Johnston, then Governor of Uganda, was intrigued by Stanley's remarks and decided to gather more information about the *atti* on his visits to the Congo. He had an unusually good opportunity at the end of 1899, when he saved a party of Pygmies from being carried off by a German showman to be exhibited as curiosities at the 1900 Paris Exhibition. They were his guests in Uganda for several months until he could take them back to their native forests. Meanwhile Sir Harry was able to question them about the horselike animal which lived in their forests.

> They at once understood what I meant; and pointing to a zebra-skin and a live mule, they informed me that the creature in question, which was called OKAPI, was like a mule with zebra stripes on it.

Okapi or *o-api* is evidently the name that Stanley mistook for *atti*.

When Sir Harry arrived in Fort Mbeni in the Congo Free State, he questioned the Belgian officers stationed there and learned that though they had never seen the animal alive, their native troops hunted it in the forest and killed it with spears. They brought back the meat to add to the fort's provisions and cut up the skin into strips to make belts and bandoliers. One of the officers said that there should be a new skin somewhere in the fort, but Sir Harry was bitterly disappointed to find that it had already been cut into pieces, and he was able to salvage only two of them.

On August 21, 1900, he wrote to Dr. Sclater of the Zoological Society of London telling of this partial success and that the precious strips of skin would be sent off at once.

> All this time [Sir Harry remarks] I was convinced that I was on the track of a species of horse; and therefore when the natives showed the tracks of a cloven-footed animal like the eland, and told us these were the foot-prints of the okapi, I disbelieved them, and imagined that we were merely following a forest-eland.

Sir Harry was very wrong to disbelieve the natives. Had Dr. Sclater known that the beast had cloven hoofs he would have been saved from making a bloomer. When he received the strips of skin he hastily christened the beast *Equus johnstoni*, though he cautiously added a question mark after the generic name *Equus*.

Meanwhile Karl Eriksson, a Swedish officer in the Belgian service, sent Sir Harry a whole skin of an okapi and two skulls. Gradually a picture of the beast was built up from the Pygmies' descriptions and the anatomical specimens. It was as big as a medium-sized horse, a little like a giant antelope in shape, but without visible horns; on the other hand it had a long tongue like an anteater's and ears as big if not bigger than a donkey's; its thighs and hindquarters were covered with stripes. Because of this last detail alone it had been taken for a new species of zebra. Actually, as Sir Harry wrote:

> Upon receiving this skin, I saw at once what the okapi was—namely, a close relation of the giraffe.

The shape of the skull, the structure of the teeth, the ruminant's cloven hoof are all evidence that it is a sort of short-necked giraffe, a creature related to its long-extinct ancestors, a cousin, no doubt, of the Helladotherium whose remains Albert Gaudry had disinterred from Miocene deposits in Greece.

Sir Harry sent his new trophies to Professor Ray Lankester in London suggesting that it should be called *Helladotherium tigrinum*. But Lankester was better versed in comparative anatomy, and came to the conclusion that the okapi was closer to the giraffe than to the Helladotherium, almost halfway between the two, and that it deserved a genus of its own no less than Sir Harry deserved the credit for discovering it. He therefore named it *Okapia johnstoni*.

As soon as the okapi's existence—which most people had doubted—was firmly established, everyone began trying to prove that it had been known for thousands of years. In 1935 R. Perret pointed out three pictures of animals very like okapis among those carved on the side of a wadi called Oued el Djerat. There is no doubt that if the animal illustrated in Fig. 5 is not an okapi it is very closely related to it.

5. This mural drawing at Oued el Djerat in the central Sahara, shows that about 3000 B.C. the Egyptians knew an animal similar to the okapi.

When Sir Harry Johnston came home from Uganda he met Stanley in England. Naturally they discussed the new animal they had both helped to discover. Stanley told him that he thought that the okapi was only one of the many animals that would eventually be discovered in the tropical rain forest and said that he had seen a huge hog 6 feet long and some antelopes of a quite unknown type.

After the harvest of new species reaped by zoologists during the last few decades the atmosphere was optimistic. The okapi had shown that the natives' tales could be trusted, and people took an eager interest in the rumors which had been current for a quarter of a century about a sort of monstrous wild boar in the Ituri forests. It was said to be a fierce beast as black as the night and as big as a rhinoceros—or, at least, as a little rhinoceros.

In 1904 Captain R. Meinertzhagen of the British East Africa Rifles happened to acquire a skin of the beast. It had come from the forests around Mount Kenya and was unfortunately in a poor condition. A little later the same officer obtained a skull of the same animal, which had been shot in the forests bordering Lake Victoria. This enabled Oldfield Thomas to give a scientific description and christen it *Hylochoerus meinertzhageni,* or "Meinertzhagen's forest hog."

It is true that it was nowhere near the size of a rhinoceros—even a small one; all the same it was the largest of wild boars, since it may reach a height of 4 feet at the shoulder and a length of over 8 feet. It is also the blackest boar ever seen and its tusks are long and massive.

This monster was no sooner added to the zoological catalogue than a similar-looking beast was rumored to live in the other end of the equatorial forest. The upcountry natives in Liberia claimed that a large black pig called *nigbve* lived in the forests of this area. As early as 1668 Dr. Dapper had noted that in the Pepper Coast, which is now Liberia,

> There are two sorts of swine, the red called *Couja,* which are the size of our own, and the black called *Couja Quinta* which are much larger and more dangerous; for they have teeth so sharp that they break everything they bite as if they were so many axes.

The *Couja* is evidently the red river hog (*Potamochoerus porcus*). As to the black pig, some suggested that it was not a giant forest hog at all, but one of those pygmy hippopotamuses reported from the same area so long before.*

Carl Hagenbeck, the great German animal dealer, decided to get to the bottom of the confusion and sent his agent, Hans Schomburgk, out to Liberia in 1909. But it was not until June 13, 1911, after months of searching, that he came upon the creature 10 yards away in the forest. It was a shiny black and did look like a big pig, but it was obviously related to a hippopotamus. Unfortunately Schomburgk—who was the first white man to see a pygmy hippopotamus in its natural surroundings—had no means ready to catch it and had scruples about shooting an animal which was thought to be extinct and was certainly very rare.

Schomburgk could do nothing. The rainy season was setting in, and he had to return to Hamburg empty-handed. But now he knew he was not hunting a myth, and by Christmas 1912 he was in Liberia again. This time he was luckier. On February 28, 1913, having made sure that the species was much less rare than he

* In spite of what immediately follows, I am by no means convinced that the *Couja Quinta* is a pygmy hippopotamus, for the giant forest hog was later found also to live in Liberia.

had thought, he shot the first specimen. The next day he managed to capture one alive and found that though its teeth looked unpleasant it was actually much easier to tame than an ordinary hippopotamus. Besides being smaller it was much less heavy in outline, had smaller jaws in proportion, and was blacker and shinier. Moreover it was a forest animal which went into the water only occasionally to drink or bathe. Leidy had been right in putting it into a new genus, *Choeropsis,* since the word *Hippopotamus,* or "river horse," does not describe its anatomy or its habits.

Five months later Hans Schomburgk confounded the skeptics by bringing back five live pygmy hippopotamuses to Hamburg. These little beasts weighed one tenth of the weight of a true hippopotamus. An adult male was no more than 2 feet 6 inches high and 5 feet 10 inches long.

It is little over half a century since the largest of the bears and the largest of the apes were discovered and the largest rhinoceros was found to exist in a country where it had been quite unknown. The largest of the lizards was not discovered until 1912, when an airman made a forced landing on Komodo, a small island between Sumbawa and Flores in the Malay archipelago. The only men who lived there were convicts deported by the Rajah of Sumbawa. The airman came back with a tale that he had met fierce and monstrous dragons, at least 12 feet long, which according to the inhabitants ate pigs, goats, and deer, and even attacked horses. Needless to say nobody believed a word of his story.

But soon afterward the existence of these giant reptiles was confirmed by Major P. A. Ouwens, curator of the Botanical Gardens at Buitenzorg. He had been corresponding with J. K. H. van Steyn van Hensbroek, the Civil Administrator of Flores, about them since December 1910. The islanders had told this official that on the neighboring island of Komodo, there was a "land crocodile." Van Steyn was interested, and when his duties took him to Komodo he learned from Kock and Aldegon, stationed on the island with the pearl-fishing fleet, that the lizards were sometimes over 20 feet long. But fishermen are proverbially unreliable judges of size.

Van Steyn was more modest. During his stay he obtained the skin of a specimen 7 feet long and sent it with a photograph to Major Ouwens, saying that he would try to catch a bigger one, but it would not be easy, since the natives were terribly afraid of its teeth and its thrashing tail.

The Zoological Museum at Buitenzorg sent a Malay animal catcher, who managed to bring back four specimens alive, the largest being 9 feet 6 inches and 7 feet 8 inches long. A little later, according to van Steyn, a Sergeant Beker shot one 12 feet long.

Major Ouwens at once recognized that these monstrous prehistoric-looking lizards were a giant species of monitor lizard which he described under the name of *Varanus komodoensis.*

When the Komodo dragon was better known it was found that it was only rarely dangerous to man. All the same it is a fearsome-looking beast which certainly kills and eats buffaloes, wild pigs, and deer. It is now to be seen in several zoos, where you have only to watch at feeding time to witness its extraordinary gluttony, which may perhaps explain why, although *Komodo* means "rat island," there is not a single rat there now.

While we are in the East I should mention in passing that in 1918 a freshwater dolphin of a quite unknown genus was discovered in Lake Tung-Ting, 650 miles from the mouth of the Yangtze River. It was all white, 8 feet long, and had a long beak.

Meanwhile there were plenty of zoological discoveries still to come from Africa. In 1908 Lydekker described a new species of spiral-horned antelope from Abyssinia, the mountain nyala (*Strepsiceros buxtoni*). More interesting from our point of view were the pygmy gorilla and pygmy chimpanzee.

The American zoologist Daniel Giraud Elliot described the pygmy gorilla in his monumental three-volume monograph on the primates in 1913. His description of *Pseudogorilla mayema* is based solely on a few skeletons and skins from the delta of the Ncomi and the Rembo south of Fernan Vaz. Even its appearance is still little known. Its coat is dark gray, except on the head and shoulders, where it is a slightly reddish chestnut. The male rarely exceeds 4 feet 6 inches—the height of a female chimpanzee—while the two subspecies of large gorilla may be as high as 6 feet and 6 feet 6 inches.

After fifty years we are still waiting for more detailed news of this "pygmy giant ape." Yet, strange to say, the existence of this species about which we know so little has never been doubted, although the pygmy elephant, which is much better known, has been so bitterly disputed that I have had to treat it as an unknown animal in a later chapter in this book.

The pygmy chimpanzee was discovered in a museum. In 1925

Ernst Schwartz examined the skins and skeletons sent to the Congo Museum at Tervueren in Belgium from the country between the left bank of the Congo River and its tributary, the Kasai, and he realized that they must be a new and quite distinct variety of chimpanzee. The pygmy species is thinner, its limbs are slenderer; its skeleton weighs only half of that of the ordinary chimpanzee, and it is shorter by 15 per cent. Its face and ears are quite black from birth, except around the mouth, while the other chimpanzees are born light and darken with age.

Later the pygmy chimpanzee was observed in its native forest, and Professor Urbain brought a live specimen to the Vincennes Zoo. Though every care was taken, it lived only for a year, from 1939 to 1940. A few specimens can now be seen in the zoo at Antwerp.

Another no less sensational discovery was made in the same way—and in the very same museum. It is true that the animal was not one of the large quadrupeds with which we have been refuting Cuvier, but merely a bird the size of a pheasant. However, since it also comes from the equatorial forests of Africa, still so full of surprises, and since the discovery of a quite unknown genus of bird is almost unheard of, the story deserves to be told here.

In 1909, while Schomburgk was on the pygmy hippopotamus's track in Liberia, an expedition was organized by the New York Zoological Society in order to bring back a live okapi to the Bronx Zoo: Herbert Lang and his young disciple Dr. James P. Chapin took part in it. The attempt failed, but when they got home they found that the tall native headdresses they had brought back as curiosities included two reddish feathers striped with black, which none of the experts could identify.

The problem remained unsolved until 1936, when Chapin happened to recognize some similar feathers on two stuffed birds, thrown out among the lumber in the Tervueren Museum. At first he did not believe his eyes, especially as one of the birds was labeled *"Pavo cristatus,* juvenile, imported." All the same it was not a common peacock, but a hitherto unknown genus of Congo peacock which he named *Afropavo congensis.* It was the first unknown genus of bird discovered for more than 40 years.

Chapin could not wait to capture a living specimen. He flew

to Stanleyville, where he found that the news of his discovery had gone before him and no less than eight specimens were waiting for him. He was astonished to learn from the natives in the forests of the Ituri (the site of so many discoveries) that though the Congo peacock had a restricted habitat it was very common in the vast stretch of country between the Ituri and the Sankuru.

This was the sixth species of largish animal to be discovered in the equatorial rain forests of Africa in less than 50 years, and it was not the last. In 1919 Glover Allen discovered in the Ituri forest a genet of aquatic habits which lives on fish. It is the size of a domestic cat with a spendid chestnut coat, marked with white spots on its face, and with a thick black bushy tail. This is all we know about it. No one has ever found any specimens since. The natives had apparently never even heard of it and so it has no local name. In English it is generally called water civet, in Latin *Osbornictis piscivora.*

The year 1929 takes us back to China, where Father David's black-and-white "bear" was on the point of being considered an extinct species when two sons of President Teddy Roosevelt—Colonel Theodore Roosevelt and his brother Kermit—saw one dozing in the top of a hollow pine tree. A salvo of shots pitched the poor brute into its last sleep, and its stuffed skin was soon being admired in the Field Museum at Chicago.

This did not prevent further slaughter in 1931, 1934, 1935, and 1936. By now it was quite clear that it was not a bear but closely related to the panda, so it came to be called the giant panda.

In 1936 William Harkness, who had gone to China to catch a giant panda, died suddenly in Shanghai before he had a chance to do so. But his widow was persuaded by Gerald Russell to carry on his task and set off on the panda's track in 1937. She knew next to nothing about the animal, and still less about China, but she had the luck she deserved, and, where the most experienced explorers had failed for 70 years, she succeeded. Within a fortnight of her arrival in the field she found a baby panda in a hollow tree. It was crying as if its heart was fit to break. She picked up this child of her husband's dreams and nursed it in her arms. It survived a delirious welcome to the United States, but in the Brookfield Zoo at Chicago it swallowed a branch too gluttonously and died in March 1938. The second time Mrs. Harkness was luckier. She brought back another female giant panda, called Mei-mei, who settled happily in her new home.

The last great zoological discovery in Asia aroused bitter controversy.

The inhabitants of Cambodia have long known that in clearings in the forests in the north of the country there is a wild ox which is either the gaur, *Bos* (*Bibos*) *gaurus,* nor the banteng, *Bos* (*Bibos*) *banteng*. It is called *kou-prey,* or gray ox.

It had been reported in 1930 and 1933, when Professor Achille Urbain, director of the Vincennes Zoo, went to Indochina in 1937. There he saw some handsome horns of this beast among the trophies of a veterinary surgeon called Dr. R. Sauvel, and persuaded him to catch a live specimen. The professor also examined an adult male which Sauvel had just shot near Chep, and he named the beast *Bos* (*Bibos*) *sauveli.*

It was after the scientific description had been published that the trouble arose. The Cambodians, when questioned, said there were *two* kinds of unknown ox in the north of the country, the *kou-prey* and the *kou-proh*. Then it turned out that the *kou-prey* was merely the *kou-proh*'s calf. A more damaging aspersion was that the *kou-prey* (or *kou-proh*) was often seen mixed with herds of banteng. Might it not therefore be a hybrid of a banteng with some other ox? Other naturalists even suggested that it was a hybrid between a half-domesticated gaur and one of the wild species.

Naturalists, it seems, will invent anything to avoid admitting that a new large animal has been discovered. None of this criticism was justified. Professor Urbain had watched a whole herd of *kou-prey*. And the minute anatomical study which Harold Jefferson Coolidge made in 1940 showed that the *kou-prey* was so different from the other Asiatic oxen that he thought it should be put in a new genus, *Novibos;* but I think this is going too far.

The last great zoological acquisition, the Andean wolf, is known only from one skin. Today it is in exactly the same position that the pygmy hippopotamus and the giant panda were in at the end of the last century, halfway between a myth and a real animal; there are palpable remains, but there is still bitter controversy about the creature's identity.

In 1926 Lorenz Hagenbeck, the famous German animal dealer who carried on his father Carl's business, happened to buy in

Buenos Aires the skin of a large canine animal which was said to come from the Andes. The skin was sent to Germany, where it was passed from museum to museum, ending up in Munich. In 1940 it came into the hands of Dr. Ingo Krumbiegel, who agreed that it was a mountain species of the maned wolf. But he did not dare to undertake a description of it on the evidence of only one skin; for with the Canidae you never know whether it may not be a mongrel offspring of some unusual cross. No zoologist could survive the blow if his new species turned out to be a cur from the gutter.

Dr. Krumbiegel shelved the matter until 1947, when he learned from Lorenz Hagenbeck that when he bought the skin he saw three others exactly similar to it. There was therefore little danger that they were hybrids, especially as the skins had so many peculiarities.

Dr. Krumbiegel then related this skin to a skull which he had examined many years before and which had come from a group of specimens collected in the Andes. At that time he had thought that it could only belong to a maned wolf from southern Brazil, Paraguay, and the Argentine (*Chrysocyon jubatus*), which the Indians call *aquara quazu* or "big fox," * although it measured a little more than 12 inches, while the average of twenty maned wolves was a little less than 10 inches. It therefore probably belonged to the same species of Andean wolf as Hagenbeck's skin. And on these two pieces of evidence Dr. Krumbiegel published a first summary description.

His unwearying attempts at reconstructing the animal eventually resulted in some excellent drawings, reproduced in Plates 1–2, following page 146, and clearly show how much the Andean wolf differs from the maned wolf of the pampas. Its blackish-brown fur is not only darker but also much thicker; the hairs on the back are 8 inches long. The legs are shorter and more solid, and the claws are more powerful. The ears are smaller and rounder; and the jaws are a little heavier and stronger. It is clear at once that one is an animal of a cold climate and that the other runs on the plains.

These differences are so marked that Dr. Krumbiegel has pro-

* The maned wolf looks like a tall fox, and I have always thought it has been mistakenly put among the dogs and wolves (*Canis*), when it is in every respect like the foxes (*Vulpes*). Dr. Frechkop tells me that he examined the specimen that recently died at Antwerp Zoo and found that its pupil was slightly oval, which confirms my suspicions and the justice of its Indian name.

visionally put the Andean wolf in a new genus, though the question will not be finally settled until we have further and more complete specimens.

But in fact the fauna of the Andes is so little known that it looks as if it will be many years before the question of the mountain wolf is solved.

There should be no need to refute Cuvier's dictum any further. The world's zoos can boast of the Indian tapir, Cotton's white rhinoceros, Grévy's zebra, and Przewalski's horse, of Schomburgk's and Père David's deer, the okapi, the water chevrotain, the pygmy hippopotamus, the Assam takin, the gerenuk, dibatag, and a host of other new deer and antelopes, the *kou-prey,* the giant panda, the gelada, both lowland and mountain gorillas, the giant siamang and the pygmy chimpanzee, the snub-nosed monkey, the Kodiak bear, the monkey-eating eagle, the Congo peacock, and even the Komodo dragon. In our museums there are stuffed skins of the giant forest hog, the pygmy gorilla, the white dolphin of Tung-Ting, the water civet, and the Andean wolf, and no doubt the skins of many unknown animals besides.

The lesson should have sunk in by now, yet there are some zoologists who will never learn. In 1934 Dr. C. Anderson of the Australian Museum could still write:

> Although there are many kinds of land animals yet to be discovered, they are mostly of small size, and it is safe to say that there is no mammal, bird, or reptile of large dimensions and unusual structure which is entirely unknown, and which would not fall naturally into some well-recognized group.

What can I say? I must leave him to be dealt with by the zoologist who in 50 years' time compiles the list of the great zoological acquisitions of the second half of the twentieth century.

SUMMARY OF GEOLOGICAL PERIODS

Era	Period	Millions of years since the beginning of the period (approximate)	Animal activity
Quaternary	Recent	—	—
	Pleistocene	1	Man appears
Tertiary	Pliocene	15	The mammals multiply and expand
	Miocene	35	
	Oligocene	45	
	Eocene	60	Rise of the mammals
Mesozoic	Cretaceous	140	Collapse of the reptile empire
	Jurassic	170	First birds, peak of the reptile empire
	Triassic	195	First mammals, decline of the amphibians
Paleozoic	Permian	220	Amphibian empire
	Carboniferous	275	First reptiles and insects
	Devonian	320	First amphibians
	Silurian	350	Fish multiply
	Ordovician	420	First fish
	Cambrian	520	Reign of the marine invertebrates
Archaean	Pre-Cambrian	2,500	Single-celled animals and sponges

CHAPTER **3**

THE SURVIVORS FROM THE PAST

She liv'd alone, and few could know
When Lucy ceased to be . . .
WILLIAM WORDSWORTH,
"She dwelt among th'untrodden ways."

WHAT IS A "LIVING FOSSIL"? The exact meaning is relevant here since we shall often be concerned with the possible survival of animals universally thought to be extinct. For it is a fact that some animals still unknown to zoologists seem to be well known to the paleontologist, though only in a fossil state. Yet many zoologists and most paleontologists steadfastly refuse to admit that they can possibly survive, and their reason for doing so is all the more obscure since the world teems with "living fossils." What then is a "living fossil"?

For the man in the street "living fossils" are extraordinary-looking and very rare creatures which survive from a vanished age. Yet, as likely as not, he washes himself with the skeleton of a sponge which was hardly altered since the Cambrian period at the beginning of the Paleozoic era, hundreds of millions of years ago. And when there is an R in the month he may well eat mussels dating from the Triassic and oysters from the Jurassic. Nor are they necessarily so rare. The horseshoe or king crab (*Limulus*) existed as a genus in the Triassic and is still so common after nearly 200,000,000 years that its crushed body is used as fertilizer. And in Lake Kyogo in the Congo there are fisheries which catch 4,000 tons of the Devonian lungfish *Protopterus* every year. Nor do you need to go to the ends of the world or dive into the sea to find a "living fossil" as old as the much-publicized coelacanth. In your cellar there are probably dozens of "living fossils" from the Devonian and Carboniferous periods spinning their webs. The tarantulas in the lower red sandstone in Aberdeenshire are very like our modern spiders; likewise spiders of a modern type have been found in seams of coal in England, Bohemia, Silesia, and Illinois. Nor need they be confined to a limited area. Australia, the kingdom of the monotremes and marsupials, is inhabited by no land animals which are not "living fossils." All the others have been imported by man.

It therefore seems better to confine "living fossils" to those creatures which are among the few survivors of otherwise extinct groups, and ought "logically" to be extinct themselves. Yet from this point of view one might reasonably say that all the amphibians and all the reptiles are "living fossils" since there are now relatively few survivors of these once-flourishing groups. The amphibians ruled the earth in the Carboniferous and Permian periods but now they consist only of inconspicuous salamanders and newts, frogs and toads, and humble caecilians without legs.

In fact it is generally agreed to limit "living fossils" to very small groups of survivors which have perpetuated themselves throughout the ages while the larger groups have been undergoing vast changes and still giving birth to new types.*

Other authors are more exclusive and include only those unique and solitary survivors of once-considerable groups; the pearly nautilus, for instance, that odd shelled cephalopod which lives 25 fathoms down in the Pacific and the Indian Ocean and, like the horseshoe crab, has remained unaltered since the Silurian period in the middle of the Primary era; or the *Latimeria,* better known as the coelacanth, that crossopterygian fish which comes straight out of the Devonian period; or, finally, the New Zealand tuatara (*Sphenodon punctatus*), the sole survivor of the order of Rhynchocephalia, which dates from the end of the Primary era, before the days of the giant dinosaurs. The tuatara is a large olive-green lizard, some 2 feet 3 inches long, which still has a vestige of a third eye on the top of its skull. This is called the pineal eye and

* The layman may be glad of a reminder of the main lines of the zoological hierarchy. The animal kingdom is divided up into phyla, each phylum into classes, each class into orders, each order into families, each family into genera, each genus into species. For instance man belongs to the phylum of vertebrates, the class of mammals, the order of primates, the family Hominidae, genus *Homo,* species *sapiens.*

These divisions are sometimes further subdivided or grouped into suborders, superfamilies, or infraclasses—and every conceivable sub-, super-, and infra- —but when carried to extremes this only creates more confusion. It is well to remember that in nature there are only individuals, and that all these man-made categories are invented to clarify our knowledge not to obscure it.

is found in the higher vertebrates in the shape of the epiphysis or pituitary gland.

A recent addition to these lone wolves of evolution is the *Neopilina galathea,* which was first described in 1957 by Dr. Henning Lemche without benefit of sensational newspaper headlines. Yet from a zoological point of view it is even more extraordinary

6. The tuatara of New Zealand is older than the great dinosaurs.

than the coelacanth. It is a small mollusk like a limpet which the Danish oceanographic vessel *Galathea* fished up off Costa Rica in 1951. When carefully examined in the laboratory it proved to be the sole representative of a whole class which had been extinct since the Paleozoic era, the Monoplacophorans. This discovery has meant that the classification of the mollusks has had to be completely rearranged.

"Living fossils" can best be defined as *stationary* species. There are quite a number of types of animal whose evolution seems to have stopped long ago, for their structure has not altered appreciably since distant ages. Admittedly it is not always easy to say for certain that an organism has not changed for millions of years, for our knowledge of past creatures is often slender and based on no more than a shell, a skeleton, or a mere impression in the mud. We cannot be sure that the flesh which has perished has not undergone any change in its physiology or cellular structure. We may also grant the name to creatures which have preserved a set of important and archaic characteristics, though they would be "living fossils" only in respect to these particular features.

The hoatzin (*Opisthocomus hoatzin*) of the Amazon is a good example of this. When fully grown it looks ordinary enough, like a sort of crested pheasant, yellowish in color with an olive back and dull red belly. But the young bird still has well-developed claws on the first two fingers of its wings, thus betraying its distant origin. When hoatzin chicks use their clawed wings to climb branches, crawl on the ground, or swim after tadpoles, they look just like little reptiles, and remind one irresistibly of the archaeop-

7. The young hoatzin still has claws on its wings.

teryx, the reptile-bird of the Jurassic, when birds had teeth. The hoatzin has also preserved other significant archaic features. It does not cry like a bird, but croaks like a frog and gives off a strong smell of musk like a crocodile and some of the turtles. To deny that it is a "living fossil" would be absurd.

The fact of the matter is that there are "living fossils" surviving from all past ages. The oldest are the radiolarians, unicellular creatures with a flinty shell convoluted like a filigree jewel, remains of which have been found in Pre-Cambrian strata; that is to say, they date from 2,000,000,000 years ago, before the beginning of the Paleozoic era. But the most recent are not the least interesting. The opossums, the very rare West Indian almique (*Solenodon*), the armadillos, the spectral tarsier, and the tapir all date from the Eocene, almost 70,000,000 years ago. The shrews, the pangolins, and many of the monkeys in the Old World are relics of the Oligocene. The hedgehogs, the okapi, and the aardvark come from the Miocene, and so on.

The point I have been trying to make in discussing all these "living fossils" is that there is no creature, however "primitive," which could not have survived until today, and that in fact we find lingering examples of most of the groups which once flourished on the earth.

Yet the mere suggestion that a dinosaur, a flying reptile, or an ape man might survive will exasperate most scientists. These creatures, they say, have been completely extinct for a long time, the argument being that their fossil remains have not been found in geological strata later than some remote period.

This is to rely unduly on the hazards of fossilization. The conditions necessary to insure that a body is fossilized, that it is preserved from scavengers of all kinds and from the destructive effects of air and water, occur only once in a thousand million times. Generally it is some exceptional accident that produces them; the animal falls into a salt lake or a frozen marsh that never thaws again, is swallowed up in mud, a lake of pitch or a puddle of resin, or is suddenly smothered by an avalanche, a stream of lava, or a sandstorm. The characteristic fossils of any geological stratum are therefore far more likely to belong to the commonest and most widespread groups of animals of the corresponding period, to the species with a virtually indestructible shell or carapace, and also to the most vulnerable species, that is to say the slowest and least intelligent. This may be why we have so little physical evidence about the past history of apes and man.

The absence of fossils of a certain type in a geological stratum may mean that the type was not common at the time, or that it had no hard parts (or no very hard parts, as in the toothless mammals), or that it was nimble and cautious enough to escape accidents that might cause fossilization, or merely that it had the good luck to escape any such accidents. It is the first alternative that interests us, for we know next to nothing about the animals of the past which were not widespread, and of which we have but a few rare fossils.

To show where the line should be drawn between a common species and one that is not widespread, I would say that the common land animals today are man and his domestic animals, a lot of rodents and bats, a good many birds, and vast numbers of insects and small mollusks. All the other animals—most of the mammals, in fact—can be considered as rare and on the road to extinction.

Moreover, as Professor Caullery has pointed out, "even those organisms which are fossilized do not all reach us, by a long way. Many of them subsequently disappear as a result of a series of accidents; strata fold and laminate, change shape, dissolve and are eroded." And he adds:

> Finally, if we consider the fossils which have actually survived until our time, we must not overlook the fact that we can reach only an extremely small proportion of them. A large part of the sedimentary formations is today submerged beneath the seas and is thus quite inaccessible to us. Of the sediments on the dry land of our continents, we can normally examine only the outcrops, that is to say an infinitesimally small part.

The late Professor Leon Bertin of the Paris Museum stated quite categorically, "In paleontology negative evidence means nothing." There are any number of groups of animals which must have existed at certain times, although not the slightest trace of them has been found in the corresponding strata. And while fossils are occasionally found where they have been deduced to be, they are more apt not to be there.

The *Latimeria chalumnae* which was fished up in a dragnet off the South African coast on December 22, 1938, is a striking proof of this. The group of Crossopterygia, which gave birth to this large fish with lobed fins at the same time as it spawned the main body of land vertebrates, had been extinct since the end of the Primary era, some 200,000,000 years ago. Remains of some members of the Coelacanth family have been found in the Jurassic, and a single one, *Macropoma,* at the end of the Mesozoic era. But for some 70,000,000 years there have been no more. Nor did the tuatara appear from nowhere, yet no trace has been found of any rhynchocephalian in strata less than 135,000,000 years old. The *Neopilina* belongs to a group which had been thought to be extinct for 280,000,000 years. There could be no better proof of the limitations of paleontology.

The refusal of many zoologists to consider that certain fossils could have survived seems to be due to notions which have long been rejected. When it was first shown that fossils were actual remains of long-dead animals, this discovery had to be made to agree with what happened in Genesis. This led to the theory that the ancient fauna had been destroyed by universal disasters, the last of which was Noah's flood. Cuvier refused to admit anything which would disagree with the letter of the Scriptures, so he invented his famous theory of the Revolutions of the Globe. Each of his supposed periodical disasters must have been followed by a new creation, at least if one is to account for the appearance of new forms—or so his pupil Alcide d'Orbigny concluded, estimating that twenty-seven successive creations were necessary to explain all the known fossils. But Cuvier said, "I do not pretend that there had to be a new creation to produce the existing species: I only say that they used not to exist in the places where they are now seen, and that they must have come there from elsewhere."

This absurd explanation implies that originally the fauna on the earth included all the species which have ever existed. After each almost universal disaster the earth would have been repopulated by the species that had escaped destruction and thus the world's fauna would have been gradually impoverished.

This theory completely disagrees with the facts of paleontology which the father of that science had taken such pains to learn. The deeper the geological strata the less varied the forms of life: there are no birds below the Jurassic, no mammals below the Triassic, no reptiles below the Carboniferous, no amphibians below the Devonian, no fish below the Ordovician, and in the Pre-Cambrian there are nothing but sponges and the most primitive invertebrate creatures. Far from becoming impoverished, the fauna has continually become richer with the passage of time.

Moreover Cuvier explicitly contradicts his own theories, for he maintained that man could not have been contemporary with the so-called antediluvian animals. But if man did not exist at the beginning of time and has not descended from more bestial creatures, where on earth did he come from? "From elsewhere," replied Cuvier, without the courage of his religious and philosophical opinions, and all the world cried Amen.

Cuvier's baneful influence was felt for a long time and had the worst consequences. In 1854 the Torquay Natural History Society refused to publish a note about some paleontological discoveries, even though they were confirmed by three reliable witnesses, because of its "lack of probability." Some tools of worked stone had been found with the remains of extinct animals.

In 1860 the Académie des Sciences at Paris used the same pretext to refuse to publish Edouard Lartet's note on the geological antiquity of the human species in western Europe. He was obliged to have it published in Switzerland and in England, where it was received with enthusiasm. Not long after the great Boucher de Perthes died, his family, who were horrified by his heretical opinions (for he too believed that man was contemporary with extinct animals), had his works withdrawn from sale and sent them to be pulped.

We may smile at this ludicrous behavior, but even today many scientific journals refuse to publish any contribution, however seriously documented, that deals with the possible survival of certain species which date from a past age.

It is not so easy to be rid of doctrines forcibly imposed. When the theory of evolution won the day, it was at first naively interpreted by minds loath to disturb the old dogmas too suddenly. There was no longer the notion of a discontinuous series of fauna, but of a continuous chain, with the corollary that every ancient species could be considered as the direct ancestors of its more recent relatives. This principle implied that the existing lizards of today were descendants of the dinosaurs of the Secondary era. Hence it followed that an animal that belonged to one age could not be found in a subsequent one, for it was constantly altering. Thus there was a new reason for automatically ruling out the survival of archaic forms.

Latterly the notion of "living fossils" has become more familiar, and the evolutionary tree is seen to be not a straight bamboo of successive stages, but a complex of almost parallel branches bursting occasionally into bushes of twigs.

A truly original type is never the outcome of a high degree of specialization. It does not spring from the end of one of the twigs of the last bush, representing the previous group to burst out in sudden expansion. It always derives from a flexible, unspecialized type on the stem of the bush. After an uneventful evolution, and inconspicuous because it is usually small, commonplace, and of unspecialized habits, the type that is trying to find its way eventually comes upon the most favorable conditions and then gives birth with almost explosive suddenness to a host of very varied forms, making a bush on the family tree. Meanwhile another humble and inconspicuous scion leaves the base of the bush, one day to burst into another new group. And so on.

Thus each new form evolves parallel to the older forms, but is a little behindhand. The various groups in the animal family tree are joined only at the base. The animal empires do not take turns at ruling the earth. They encroach upon one another and destroy one another when they are both competing for the same habitat. "Living fossils" are creatures that have escaped these massacres, either because they have never found opponents of their own size, or, as more often happens, because their territory has not yet been disputed or has only just begun to be conquered.

For this reason we can expect to find them chiefly in the remotest corners of the earth: islands, high mountains, underground caves, virgin forests, swamps, and deserts.

In the previous two chapters I have shown that the earth may still hold plenty of large unknown animals and that in fact the zoological catalogue is constantly being added to. We now know that among the future zoological discoveries there may well be animals which are known but thought to be extinct.

The history of the discovery of an unknown animal almost always follows the same pattern. At first it is utterly unknown to the Western world. Then, by questioning the natives, travelers or merchants first come to hear of its existence. There is not a single example of a large animal which has remained quite unnoticed by the people who live nearest its habitat.* The local descriptions of the beast are often only roughly accurate and exaggerated in some respects. Nevertheless the different versions of the story, even those told by tribes far removed from one another, agree so well that they eventually arouse the interest of a few naturalists. Finally white hunters' or prospectors' tales confirm the native legends. But not until the animal is brought back, dead or alive, do the skeptics admit the evidence. There is only one variation in this pattern. Sometimes the animal described by the natives is thought to be a fossil by the Western world, and its bones, or even its mummified hide, appears in paleontological collections.

If we want to have an idea of the still unknown animals that may be discovered in the future, our best plan is determined by this fixed pattern of events. First we should listen carefully to the native tales, picking out the details that agree in accounts from different sources. Then we should try to see whether this description fits any species in the local fossil or subfossil fauna, that is to say, one that is thought to be recently extinct. Actually, when it is a species that is supposedly very recently extinct we shall have to be particularly careful since it is possible that the animal's description has been handed down by tradition from ancient times. All the same we should remember that Van Gennep, who has made a very thorough study of the process by which legends are formed,

* The water civet, though small and amphibious, and therefore inconspicuous, may perhaps be taken as the exception which proves this rule.

considers that "the memory of a historical fact does not continue among communities *which do not use a written language* for more than five or six generations, that is to say for 150 years on the average, and 200 years at the most." It therefore takes only two centuries to corrupt or wipe out the memory of historical facts or events among peoples whose traditions are purely oral. This knowledge will be most useful to us.

The reader may be surprised at the trust I generally put in the evidence of what are somewhat condescendingly called "primitive" peoples. But because their conception of the world is utterly different from ours there is no need to say that they are all liars.

Westerners are much more severe on the metaphors and similes of primitive peoples than they are on their fellow countrymen's tales. They do not mind if you say that after walking through a fog you could "cut with a knife" you saw an animal with an "interminable" tail which made off "like lightning." But if some Australian aborigine tells them of a "devilish worm" that "spits fire" and which it is "death to look at," they shrug their shoulders and will have no more of it, when they ought to realize that this merely means that the aborigines know of an animal that is wormlike (or snakelike, it comes to the same thing), has a forked tongue like a flame, and which it is wise to avoid as its bite is fatal.

PART TWO

THE MAN-FACED ANIMALS OF SOUTHEAST ASIA

And there be mungrell and ambiguous shapes,
between a humane and a brutish Nature.
MONTAIGNE, *An Apology of Raymond Sebond.*

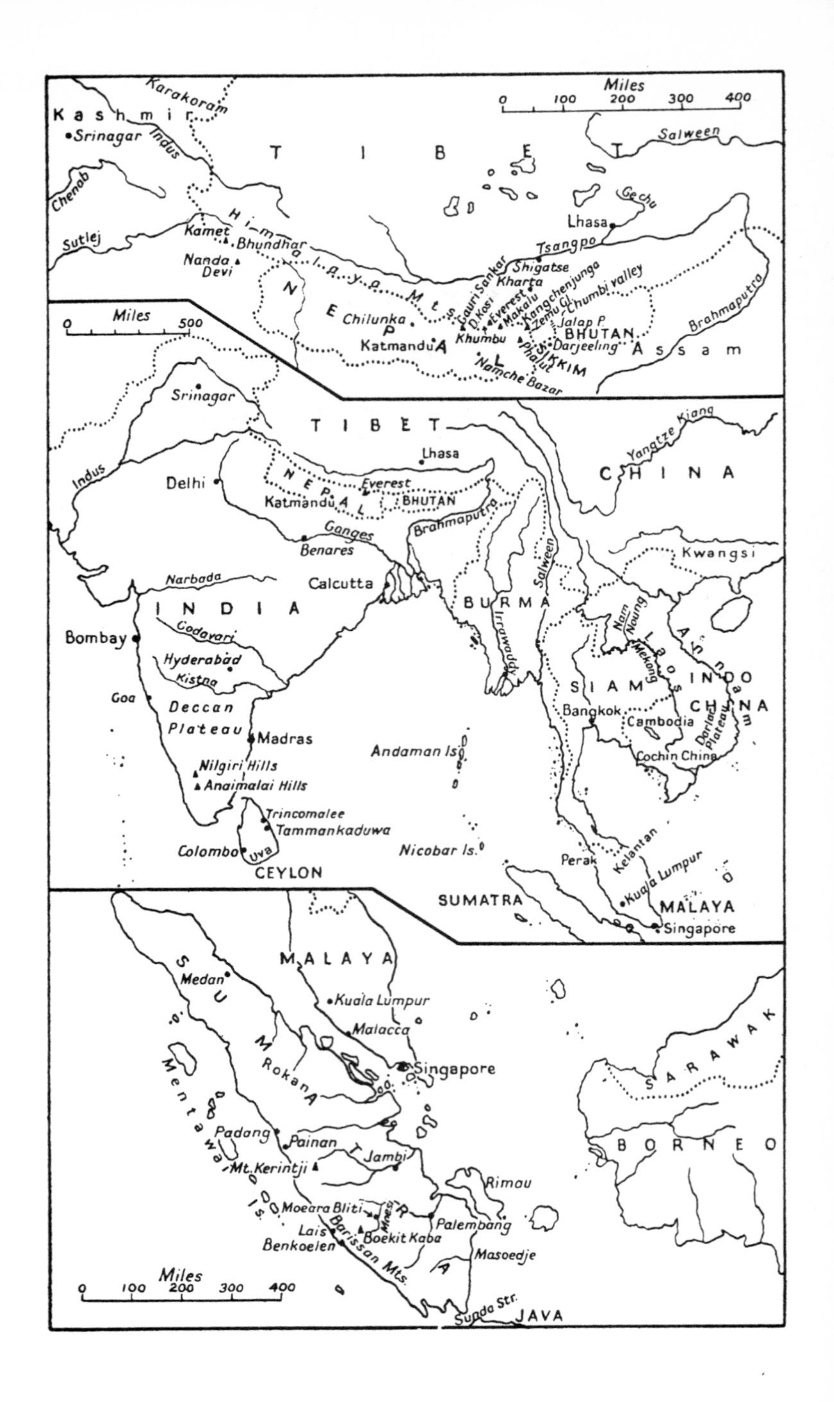

Karakoram
Kashmir
Srinagar
Indus
TIBET
Miles
0
100
200
300
400
Salween
Chenab
Gechu
Lhasa
Sutlej
Kamet
Bhundhar
Himalaya Mts.
Tsangpo
Nanda Devi
Shigatse
Kharta
Gauri Sankar
D. Kosi
Everest
Makalu
Kangchenjunga
Zemu Gl.
Chumbi valley
Jalap P.
NEPAL
Chilunka
Katmandu
Khumbu
BHUTAN
Brahmaputra
Darjeeling
Phalut
SIKKIM
Namche Bazar
Assam
Miles
0
500
Srinagar
TIBET
Lhasa
Yangtze Kiang
CHINA
Indus
Delhi
NEPAL
Everest
Katmandu
BHUTAN
Brahmaputra
Ganges
Benares
Salween
Kwangsi
Narbada
Calcutta
INDIA
BURMA
Irrawaddy
Nam Noung
Godavari
Bombay
Laos
Mekong
Annam
Hyderabad
Kistna
SIAM
INDO CHINA
Goa
Deccan Plateau
Bangkok
Cambodia
Darlac Plateau
Madras
Andaman Is.
Cochin China
Nilgiri Hills
Anaimalai Hills
Trincomalee
Tammankaduwa
Kelantan
Colombo
Uva
Nicobar Is.
Perak
Kuala Lumpur
CEYLON
SUMATRA
MALAYA
Singapore
SUMATRA
Medan
MALAYA
Kuala Lumpur
Malacca
Mentawai Is.
Rokan
Singapore
SARAWAK
Padang
Painan
Jambi
BORNEO
Mt. Kerintji
Rimau
Moeara Bliti
Musi
Palembang
Lais
Boekit Kaba
Benkoelen
Barissan Mts.
Masoedje
Miles
0
100
200
300
400
Sunda Str.
JAVA

CHAPTER 4

NITTAEWO, THE LOST PEOPLE OF CEYLON

> . . . that Pigmean Race
> Beyond the *Indian* Mount, or Faerie Elves,
> Whose midnight Revels, by a Forrest side.
> Or Fountain some belated Peasant sees,
> Or dreams he sees.
>
> JOHN MILTON, *Paradise Lost.*

IT MAY SEEM ODD that the first unknown animals to be investigated should be described as "people." But from India to the Malay Archipelago we shall constantly run up against the same question about the mysterious creatures reported to live there: are they men or beasts?

Asia may still hide unknown apes whose mental development is higher than that of the anthropoid apes and thus comes close to our own. Or it may be inhabited by men more primitive than the Australian aborigines, the Veddahs, or the African Bushmen, and still at the Neanderthal stage. Or yet again, a few survivors of those strange ape men, the Java Man and the Peking Man, may linger on today not far from the places where their bones were found.

The first place in which "beast men" are reported to have survived into historic times is India. Pliny the Elder wrote at the very beginning of the Christian era:

> *Duris* maketh report, That certaine Indians engender with beasts, of which generation are bred certaine monstrous mungrels, halfe beasts and halfe men.

We can pass over the fanciful and salacious theory of their origin for the moment and still admit the report of monstrous half-men and half-beasts. They need not have been crossbreeds.

The Greek historian Duris of Samos also mentions a whole race of "human beasts" on the banks of the Ganges called Calinges and all born with tails.

Let us look at the sources on which the account quoted by Pliny is certainly based: Ctesias and Megasthenes of the fourth and third centuries B.C. They were, as far as we know, the first Western writers to speak of the Indies.

Ctesias, a Greek, was physician to Artaxerxes II, King of Persia. When he returned from the East he wrote a book about Persia, and then a second about India, although he had not been there. It was based entirely upon rumors that he heard in Persia and on reports from travelers. His original work is lost, but fortunately many extracts appear in the *Myriobiblon,* or Book of Wonders, of Photius, Patriarch of Constantinople in the ninth century. It includes the first known mention of Pygmies in the East, whose existence was not admitted by science until 1887 as a result of work by the great French anthropologist Quatrefages de Bréau.

> In the middle of India there are black men called Pygmies. They speak the same language as the Indians and are very small. The largest are only two cubits high; most of them are only one and a half. Their hair is very long; it reaches to their knees and even further. They have a bigger beard than any other men; when it is fully grown they cease to wear clothing, but bind it round with a girdle and so make it serve for a garment. They are snub-nosed and ugly.

This description, though far from exact in its details, would cause no surprise today. The existence of Oriental Pygmies or Negritos is firmly established. But the Negritos are much larger than Ctesias's Pygmies. 1½ to 2 cubits is about 2 feet 6 inches to 3 feet. In fact the smallest men known in Asia, the Tapiros of New Guinea, discovered only in 1910, vary between 4 and 5 feet high, most of them being about 4 feet 9 inches; and these are not true Negritos; they are more like Pygmy Papuans.

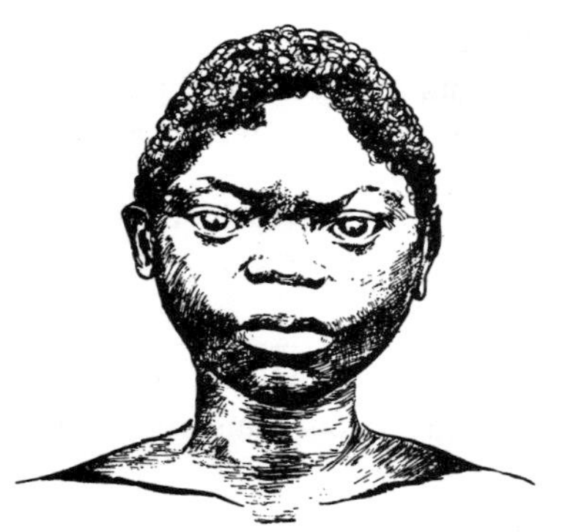

8. Typical Negrito, an Aeta from the Philippines.

The Negritos vary in color from dark chocolate to sooty black. The hair on their heads is short and woolly in tight curls like astrakhan. The rest of their bodies is practically hairless. The skull is slightly brachycephalic, face very round, forehead narrow and rounded, and chin receding. The lips are fairly thick. The rounded fleshy part of the upper lip is very characteristic. The nose is broad and flattened with very distended nostrils. Their limbs are well proportioned, but their feet are large and clumsy and turned inward a little. The Aetas exaggerate this with a very splayed big toe.

Ctesias's description hardly seems to fit the Negritos. There are no Pygmies of this race in India now, and their short and curly hair does not agree with the long mane which Ctesias gives his Pygmies. There is another much better identification which I shall come to soon.

9. Typical Veddah of Ceylon.

Ctesias and Pliny may seem unreliable witnesses, but Megasthenes, an Ionian, must have been to the banks of the Ganges when ambassador from Seleucus Nicator to the court of King Chandragupta. One is therefore all the more surprised to find that he merely quotes Ctesias. But even this is not certain, since his work is known only by quoted fragments, and when Pliny reports what Ctesias and Megasthenes have said of the wonders of India it is sometimes difficult to tell who said what.

The problem of the Indian "beast men" seemed to be solved in the year 400 when Bishop Palladius mentioned a race of strange Pygmies in Ceylon, the Veddahs, one of the most primitive peoples in the world.

The Veddahs, who long ago took refuge in the mountains and forests of the eastern part of the island, are not strictly speaking

Pygmies, for their average height is not less than 5 feet. But they are barely an inch outside the limit. They are very different from the Negritos. Their skull is clearly dolichocephalic, their hair is long and wavy. Their skin is often lighter than the Negritos', though it is still very dark. They are not at all Negroid. Their lips are thin, their nostrils wide beneath a slender nose which is depressed a little at the root. Their most remarkable feature is their receding forehead and strikingly heavy brow, beneath which their eyes seem very deep-set. Their beards are weak, and they have hardly any hair on their bodies.

Like the Negritos they are timid creatures, who run away from strangers. They live in separate families in caves or crude huts made of branches. Before the Singhalese arrived they were still in the Stone Age, but since that time they have used iron-headed arrows.

There is little doubt that the Veddahs are Ctesias's long-haired Pygmies. Only the enormous beard and small size seem to be exaggerated—and there is also the long hairy tail. Possibly the hill tribes have been confused with the wanderoo or lion-tailed monkey (*Vetulus silenus*), a black macaque with a large gray mane like a prophet's hair and beard.

In India and the Malay peninsula there are also tribes similar to the Veddahs. They have been called Pre-Dravidians, because they occupied the whole of India during the Ice Age before it was invaded by the black Dravidians, who in their turn were invaded in about 1500 B.C. by the lighter-skinned Aryans. The yellow hordes of Genghis Khan broke over India in the thirteenth century, Tamburlane came in the fourteenth century, and then Baber and Akbar in the sixteenth. These repeated tides have filled India with a mixture of many different races, and thus perhaps instigated the caste system. The Veddahs were the poor relations in this motley world and were driven into the hills and cut off in Ceylon.

The great traveler Ibn Batuta landed in Ceylon in the fourteenth century and wrote of the monkeys there:

> animals which are in great numbers in the mountains of these parts. These monkeys are black, and have long tails: the beard of the males is like that of a man.

This description probably refers to the purple-faced langur. Its fur varies from silvery gray to jet black, and it has flowing white whiskers and a very long tail. But Ibn Batuta goes on:

10. The wanderoo or lion-tailed monkey.

I was told by the Sheikh Othmān and his son, two pious and credible persons, that the monkeys have a leader, whom they follow as if he were their king. About his head is tied a turban composed of the leaves of trees; and he reclines upon a staff. At his right and left hand are four monkeys, with rods in their hands, all of which stand at his head whenever the leading monkey sits. His wives and children are daily brought in on these occasions, who sit down before him; then comes a number of monkeys, which sit and form a sort of assembly about him. One of the four monkeys then addresses them, and they disperse. After this each of them comes with a nut, a lemon, or some of the mountain fruit, which he throws down before the leader. He then eats, together with his wives, children, and the four principal monkeys; they then all disperse. One of the Jogees also told me, that he once saw the four monkeys standing in the presence of the leader, and beating another monkey with rods; after this they plucked off all his hair.

These monkeys can hardly be langurs. They seem more like Pliny's "beast men."

In 1887 Hugh Nevill, a British explorer, was told a strange story by a Singhalese hunter about a people called the *nittaewo* who agreed remarkably well with Pliny's description. The hunter had heard the story from a close friend, one of the last Veddahs in the Leanama region in the southeast of Ceylon. This old Veddah had been told by a relative called Koraleya a great deal about the *nittaewo* and their final extermination. They were a race of Pygmies who inhabited the almost inaccessible mountains in the Leanama area which lay in the south of the Veddah country.

If the *nittaewo* seemed savage Pygmies to the Veddahs, who were themselves extremely primitive and quite small, they must have been remarkable creatures. They were said to be perfect human beings, but even smaller than the Pygmies, being between 3 and 4 feet high; the females, as usual, were rather smaller. They walked upright and had no tail. They were hairy and usually reddish and they had remarkably short and strong arms. Their hands were also short, with long claws. They had no articulate language, and spoke with a sort of burbling, or birds' twittering understood only by a few Veddahs. They lived in small parties, sleeping in caves or on platforms of branches in trees covered with a roof of leaves. They fed on what they could catch: squirrels, small deer, tortoises, and even lizards and crocodiles. They did not use weapons. They ripped open their prey with their long claws and gluttonously devoured their entrails.

The Veddahs and the *nittaewo* were constant enemies. The little men had no defense against the Veddahs' bows and arrows, but they made up for it by mischief and cunning. When they found a Veddah asleep they would disembowel him with their claws. But the stronger and better-armed race won in the end. In the late eighteenth century, or so it would seem, the last *nittaewo* were rounded up by the Veddahs of Leanama and driven into a cave. The Veddahs then heaped brushwood in front of the entrance and set fire to it. The bonfire burned for three days, and the trapped *nittaewo* were all suffocated.

Unfortunately all record of the cave's position was lost when the Veddahs of Leanama became extinct themselves a few generations later. It was the last of them who told Hugh Nevill's informant. We know that he got the story from his relative Koraleya, who heard tell of the end of the *nittaewo* when he was very young.

Koraleya died in about 1870, which means that it must have been around 1800.

Hugh Nevill's report is, of course, based on information at fourth hand. One of the four links of the chain could have invented the whole story. But it was decisively corroborated at the beginning of this century when Frederick Lewis, who had never heard of Nevill's stories, explored eastern Uva and the Panawa Pattu district. From an old Veddah, who had adopted the Singhalese name of Dissan Hamy, Lewis obtained a mass of information about the habits and customs of the Veddah people, and also of the Pygmies, called "*Nittawo,*" who used to harass them. According to this information the *nittaewo* had not been exterminated more than five generations earlier—Dissan Hamy's grandfather had taken an active part in burning out one of their encampments—and their appearance and habits agreed very closely with Nevill's version.

> I may here mention [remarked Frederick Lewis] that as I was sceptical as to the story about this race, I took particular care to make inquiries at the distant village of Waradeniyama . . . when Dissan Hamy was not present, and it would have been impossible for my questions to have been anticipated. To my surprise, a very old man of the village completely confirmed in detail Dissan Hamy's description.
>
> I also made further inquiry at another village from a headman, and he repeated the same story, adding that the Nittawo were destroyed by the Veddas out of fear.
>
> It is difficult to reject as false a story, told devoid of the usual fantastic embellishments that characterize the history of mythical creatures, such as Yakko, when it is completely confirmed by parties ignorant of what the others have said. Even though tangible evidence is not forthcoming and it would be difficult to find any, of a people so primitive as these creatures appear to have been, I see no valid reason for disbelieving the statements made to me.

Accounts of the *nittaewo* by the Singhalese, or even by the Veddahs, may well have given rise to ancient legends about "beast men" in Ceylon. Frederick Lewis has shown that certain places on the southeast coast were used as ports of call by Western navigators several centuries before Christ.

The existence of the *nittaewo* seems to be established without much doubt. It remains to be discovered what they were. The eminent expert on the primates, Professor W. C. Osman Hill, has set about this task with much patience and perspicacity. First, what does the name *nittaewo* mean? Nevill derives it from *nishâda,* the name given by the Aryan invaders to the more primitive tribes. But, Professor Hill writes:

> This suggests, quite apart from other evidence, that the Veddahs themselves are not synonymous with the *Nittaewo,* even though their mainland relatives would be included under the category of *Nishâda* and they themselves would also have been, had the invaders reached Ceylon.

But might not the *nittaewo* be Negritos, who are in fact slightly smaller than the Veddahs, though only by about 2 inches, and are in some respects even more primitive? India and Ceylon could easily have been originally inhabited by Negritos of whom the *nittaewo* were the last survivors.* But in fact the Negritos do not look in the least like the description of the *nittaewo.*

If the *nittaewo* are not the most primitive of men perhaps they may be anthropoid apes, the most advanced of the monkeys. When Jacob Bontius first described the orangutan he added a legend that "the Javanese say that they are born of Indian women who couple with monkeys under the influence of a vile sensuality." This shows that the tales Pliny reported were also told of anthropoid apes. According to Nevill the *nittaewo* has been compared to the orangutan in at least one report. But this ape (see Fig. 32, p. 93) is too large, too heavy, too strictly vegetarian and arboreal, and too solitary in its habits to agree with the description of the *nittaewo.*

Actually the gibbon is a much better candidate. It is small: the largest species, the siamang of Sumatra (*Symphalangus syndactylus*), is 3 feet high when it stands upright—the same height as the *nittaewo.* It habitually walks upright on its hind legs only, and it is the only ape that does so. It lives in troops. And it is not entirely vegetarian; besides insects it will eat birds and swallow their eggs. All this is a long way from the savage and bloody habits of the

* As Aubrey Weinman has pointed out, there even seems to be concrete evidence of this:

> Most important evidence that pygmies must have lived in Ceylon in the prehistoric era is to be found in the small but extremely valuable collection of stone implements in the Colombo Museum . . . there is a whole sequence ranging from bigger to smaller ones, and the latter are so minute in structure that they could only have been used by a race of pygmies who inhabited the island in the dim and distant past.

But I do not think that he is right in considering this a proof of the past existence of the *nittaewo.* These minute tools are evidence of true Pygmies, not of the *nittaewo,* who, according to all traditions, did not use implements.

nittaewo, but the Veddahs may well have blackened the character of their greatest enemy, and according to Professor Hill a giant gibbon would make a very plausible *nittaewo*. Gibbons, however, are only found east of the Ganges and south of the Brahmaputra. It would be surprising to find them in Ceylon. But, of course, this is not a conclusive argument.

11. Gibbon walking on its hind legs.

It has also been suggested that they were bears. Bears sometimes stand upright on their hind legs and leave footprints very like those of human feet. But the only bear known in Ceylon, the Ceylon sloth bear (*Melursus ursinus*), rarely stands upright, its fur is black

—though it sometimes turns red on the surface—it is not gregarious, and is carnivorous only in exceptional cases.

The most likely theory is still that the *nittaewo* were true ape men, for the discovery of bones of the Pithecanthropus in Java and the very similar Sinanthropus in China shows that ape men (the morphological link between the anthropoid apes and man) once

12. The Ceylon sloth bear (*Melursus ursinus*).

inhabited a large part of Asia and that they were no doubt driven into the Malay Archipelago by more human invaders. They could have reached Ceylon when it was still attached to the mainland, and have then survived until very recently. It is clear enough that true ape men (which according to some authors may have been giant gibbons that had given up living in trees) fit exactly with the Veddahs' description of the *nittaewo*.

I cannot quite understand why Professor Hill thinks that their small size helps to identify them with the Pithecanthropus—which was certainly much nearer our own. But it is very likely that if the Pithecanthropus was cut off in Ceylon it would develop into a pygmy race, as isolated species so often do.

But, you may say, since the *nittaewo* are extinct, we shall never be able to prove the truth of this theory.

This is not quite true. Systematic excavation in the Leanama region might bring to light bones of the *nittaewo* in a relatively fresh condition, and thus clinch the problem.

But surely there is no hope of ever examining one of these strange ape men in flesh and blood?

Even this is not quite certain. For if the *nittaewo* are a vanished race in Ceylon there are other places in southeast Asia where hairy pygmies or "beast men" are believed to exist.

Henri Maître has shown how widespread such rumors were among the Moi in Indochina at the beginning of the century:

> The "wild men" of Nam-Noung are small—less than five feet high—covered with a thick coat of reddish hair, and arms or legs which they cannot bend at all. The back of the forearm has a sharp membrane like the blade of a knife which they use for cutting away the bush to make a path through the forest. They cannot climb up trees, since they have neither knees nor elbows, so they sleep leaning against tree-trunks. They live on roots and stalks of plants, and do not know how to build huts, living the same wandering life as other beasts of the forest. The villagers used to run them down and eat them, but the "wild men" have become scarcer and are no longer found. But one sometimes still comes upon the footprints that they leave behind them, like those of other men, but smaller.

The two striking details about these "wild men" are the tail and the strange anatomy of their forearms. Otherwise these russet pygmies agree well enough with the description of the *nittaewo*.

As to the tail, it seems to me significant that the Annamites used to allege that the Moi themselves had tails. The Moi are a mysterious race, of medium size (average height 5 feet 2 inches), with reddish tanned skins, the color of a sunburned European. They do not have slit eyes or prominent cheekbones. Their noses are rather straight and they sometimes have fair or even reddish hair. When the Moi became better known it was seen that they had tails no oftener than any other people,* on the contrary they attributed them to another race, more savage than themselves.

But this does not mean that the "wild men" of Nam-Noung really

* At the age of five weeks the human embryo normally has a well-formed caudal appendage. While an adult has 34 vertebrae, the embryo at this age has 38. Four or five at the end are later joined together to form the coccyx. But occasionally, by some freak of atavism, these vertebrae remain separate or even multiply to form a proper tail which may be as much as 4 or 5 inches long. Sometimes it is merely a fleshy appendage without bones, but when it is properly formed, with muscles and nerves, it can be moved like an animal's. It is usually bare but sometimes covered with hair. This malformation does not seem to be hereditary, nor is it commoner in some races than others, and it certainly could not be a characteristic of an entire people.

have tails. A tail is simply a symbol of bestiality. It is not without reason that we give the Devil a tail. It is the Mark of the Beast. Most of the ancient travelers gave tails to the savages they found, especially on islands. Marco Polo mentions men with tails in Sumatra. Gemelli-Careri finds them in Luzon, Jean Struys in Formosa, the Jesuit missionaries in Mindoro near Manila, Köping, a Swede, in the Nicobar Islands. And when Turner visited Tibet he was told of a race of men with tails living in the mountains. The tail was so stiff that they had to make a hole in the ground for it before they could sit down.

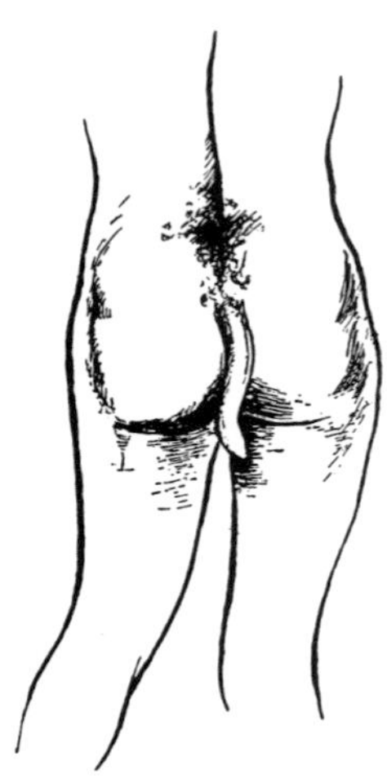

13. Man with a tail as a result of a rare atavistic freak.

Even in Europe in the seventeenth century the Spaniards believed Jews had tails, in France the people of Béarn attributed them to the Cagots who lived at the foot of the Pyrenees, and in England the Devonians believed the same slander about their Cornish neighbors.

The legends of the whole Indo-Malayan region show that a great confusion reigns in the native mind between human races who live in the forest and the local apes, the orangutans, and various gibbons. This may be partly explained by the numerous successive invasions, both of animals and men, which this country has undergone. Some relics of these occupations have survived here and there on islands, in inaccessible mountains, and in jungles that are hard to penetrate.

In the Miocene period there were gibbons (*Pliopithecus*) even in Europe; in the Pliocene they came no farther west than Egypt (*Prohylobates*); and now they are found only in Asia east of the Ganges. Up till the Pleistocene period there were still orangutans in China; now they are confined to Borneo and Sumatra. Much the same thing seems to have happened to the Pithecanthropus except that none is supposed to have survived. Then came the tiny Negritos, who once occupied all Burma and Indochina, and now survive only in the Malay peninsula, the Andaman Islands, and the Philippines. They were followed by the little Veddahs, whose vast empire once stretched across India, but who now consist only of a few hill tribes on the Malabar coast, and have fled to the islands, where they are gradually dying out in their forest retreats in Ceylon, Sumatra, and the Celebes, as well as in Malaya. All these little creatures have been beaten, massacred, and driven out by larger and better-armed invaders. The struggle for existence is essentially fratricidal. Your bitterest enemies are those most like yourself and who want the things that you do.

Thus, in a series of tidal waves, India passed from the reign of the monkeys to the reign of modern man. The succession was not that of our family tree, for, though related, the invaders were not directly descended from one another. Nor was it a regular progress toward man. No doubt there were ups and downs. The battle went to the strongest, and the strongest is not necessarily the most intelligent. A race of gregarious monkeys—like the baboons—could easily conquer a country occupied by men living in scattered families. And even today there are places where man cannot conquer the insects.

But in the end the battle would always be won by a combination of intelligence, brute strength, ferocity, and force of numbers.

The problem, then, is: to which of these waves of invaders did the *nittaewo* and the "wild men" belong? These hairy pygmies seem to have preceded the Negritos and the Veddahs. But many of their habits and traits are not those of even the highest apes which lived there before the first men. Can they belong to the race of the Pithecanthropus which no doubt occupied the whole of southeast Asia at the end of the Pliocene period—that is to say in an intermediary age? They are too small to be the *Pithecanthropus erectus* of Java or the *Sinanthropus pekinensis* of China, which were the size of modern man. The Dutch paleontologist Von Koenigswald discovered in Java in 1939 and 1941 remains of the skulls of even larger ape men which he called *Pithecanthropus robustus* and *Meganthropus palaeojavanicus*. If these creatures were proportioned like men—and this is pure surmise—the first would have been 6 feet 6 inches high and the second between 8 and 10 feet. Thus there are at all events considerable variations in the size of the Pithecanthropus. It would not be at all odd if we also found a pygmy Pithecanthropus like the African Australopithecus. And if the *nittaewo* and their Malay and Indochinese brothers really are creatures between men and apes, they must certainly be pygmy Pithecanthropi.

All this is based on a very slender scaffold of hypothesis. However, Professor Hill, who believes the *nittaewo* to be a Pithecanthropus, points out that in Sumatra there is a legend about an almost identical creature called the *orang pendek*. And this time it is not a case of vague and fantastic tales among primitive people —like the reports of the forest demons with cutting arms. The *orang pendek* has been talked of since time immemorial. It has been seen by educated Malays and even by whites, and its description is therefore quite matter-of-fact. We shall pursue it more closely in the next chapter.

CHAPTER 5

ORANG PENDEK, THE APE MAN OF SUMATRA

. . . or Pan himself,
The simple shepherd's awe-inspiring God.
WILLIAM WORDSWORTH, *The Excursion.*

THE NATIVES OF SUMATRA believed in an ape man long before remains of a Pithecanthropus were discovered in the neighboring island of Java, and indeed long before Ernst Haeckel had even invented the idea of such creatures in 1865.

In *The Origin of Species*, Charles Darwin did not carry his theory of evolution to its conclusions about the descent of man lest he should shock religious opinion. His German colleague had no such qualms, and coolly put forward a complete family tree showing how man had risen step by step from the animal kingdom. As imaginative as he was learned, he even invented new creatures to fill the gaps in the genealogy. The nearest direct ancestor of man that he could suggest was the Dryopithecus, an anthropoid ape of the Miocene period, but it seemed too large a jump between this beast and Pleistocene man, so he invented a quite hypothetical creature halfway between the two, which must have lived, he said, in the intervening Pliocene period. He called it *Pithecanthropus*, or "ape man."

The existence of this imaginary beast was so well rooted in the minds of Darwinians and Haeckelians that on his sixtieth birthday they gave Haeckel a painting by Gabriel Max of a family of ape men. When this present was offered to Haeckel no one in the West knew that remains of an actual ape man had been discovered in Java by one of his disciples. A young Dutch doctor called Eugène Dubois had dug out of the Pliocene volcanic tufa a fragment of a skull with heavy brows and a receding forehead. Its cranial capacity seemed to be between that of an anthropoid ape and a man. There

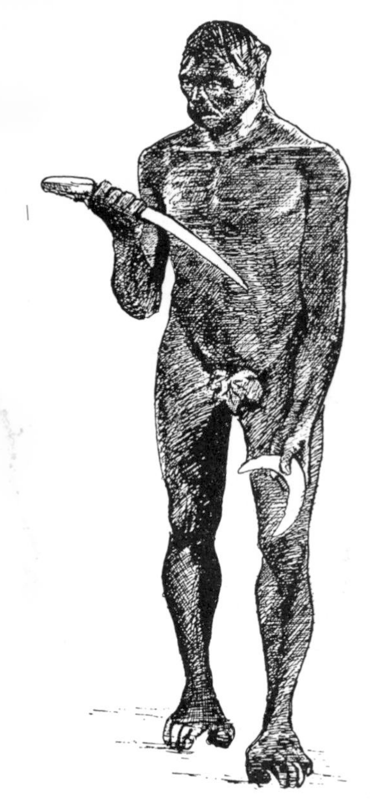

14. Reconstruction of the Java ape man in the Dutch East Indies pavilion at the 1900 Exhibition. (The fig leaf was added in the photograph on which this drawing is based.)

were also two molars and a premolar, as well as a human-looking femur which indicated that the creature walked upright. On the basis of these remains Dr. Dubois announced in 1894 that he had discovered *Pithecanthropus erectus,* the "Missing Link." A systematic excavation in Java from 1936 to 1939 organized by the Dutch paleontologist Ralph von Koenigswald brought to light three more fragmentary skulls of ape men. The discovery at Peking of rather more complete remains of a similar ape man (which has been named *Sinanthropus*, but should more logically have been *Pithecanthropus pekinensis*) has increased our knowledge of these creatures. Were they really man's ancestors? It is hard to maintain this view. Most anthropologists see them as a branch which split from the human family tree at the end of the Tertiary era. But others think they were giant gibbons which had given up living in trees; and this was Dr. Dubois's eventual opinion.

At all events they existed in Java. And across the Sunda Straits in southern Sumatra apelike men have been believed to exist since time immemorial. Marco Polo was the first to bring this story to the West, writing of Lambri (probably the province of Jambi) he says:

> In this kingdom are found men with tails, a span in length, like those of the dog, but not covered with hair. The greater number of them are formed in this manner, but they dwell in the mountains, and do not inhabit towns.

William Marsden, who was Secretary at the Residence at Benkoelen in Sumatra and edited an English edition of Marco Polo in 1818, thought that this fable arose from the actual existence of two types of native who lived in the woods and avoided all contact with the other inhabitants. They were known as *orang kubu* and *orang gugu*. The first were fairly numerous. They had their own language and ate anything: deer, elephants, rhinoceroses, dogs, snakes, and monkeys. The *gugu* were much rarer, and were covered with long hair. The *orang kubu* are very well known today and are a tribe of natives who live in the mountainous forests in the southeast of the island, exactly where Marsden put them. The *orang gugu* seem at first to be orangutans. But neither of them has a tail. A tail, however, is merely a symbol of savagery. Besides, Marco Polo had certainly not seen these appendages himself; he was trusting to a legend which already existed.

More recent stories, which are still current today, say nothing about men with tails in Sumatra: on the contrary, they tell of little wild creatures which walk upright like men and are hairy like apes, like the *nittaewo* of Ceylon. The Dutch settlers call them either *orang pendek*, which means "little man," or *orang letjo*—"gibbering man."

The natives insist that the *orang pendek* is not any of the three species of gibbon to be found in Sumatra, nor is it an orangutan, though people ignorant of the Malay language have often been misled into thinking that the natives believe in several species of orangutan. Actually *orang* merely means "man"—or "manlike creature," whether it is used in such phrases as *orang pendek* (little man), *orang malayu* (Malay man) or *orang utan* (man of the woods). (Incidentally *orang utang*—a frequent misspelling—means "man in debt"!) The Malays also use the word *orang* for non-anthropoid apes. In Borneo the famous proboscis monkey (*Nasalis larvatus*), the male of which has a nose 2¾ inches long, is called *orang blanda*, which means "Dutchman." The illustration, and the fact that the Malays themselves have rather short noses, may explain the reason why.

The *orang pendek*, according to reports, is a very shy biped which speaks an unintelligible language. It is between 2 feet 6 inches and 5 feet high. Its skin is pinkish brown and, according to most versions, covered all over with short dark brown to black hair. It has a head of jet-black hair forming a bushy mane down its back. It has no visible tail. Its arms are not as long as an anthropoid ape's. It hardly ever climbs in trees, but walks on the ground. It is supposed to walk with its feet reversed, the heels facing

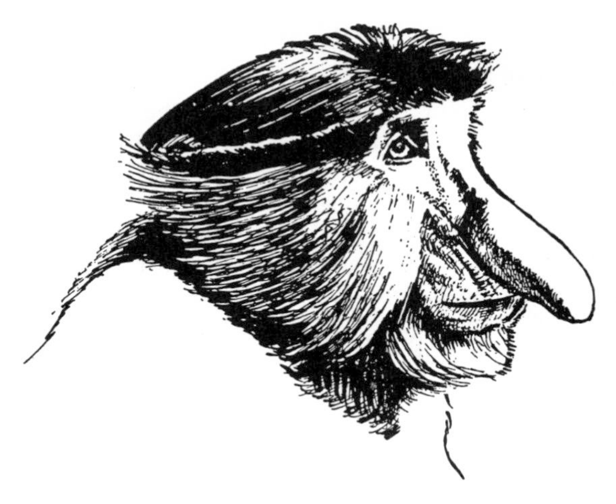

15. The proboscis monkey of Borneo, which the natives call *orang blanda* or "Dutchman."

forward. Otherwise its habits are very ordinary. It eats young shoots, fruit, fresh-water mollusks, snakes, and worms, which it finds by turning over stones and even, with herculean strength, the trunks of fallen trees. It is very partial to durian fruit. And sometimes it raids banana or sugar-cane plantations, or the natives' gardens. This man-faced beast is known all over southern Sumatra below the equator. As Marsden's note shows, Western travelers had heard of it at the beginning of the nineteenth century, but it was not for another hundred years that the Dutch settlers began seriously to believe that it existed.

In 1917 it was mentioned in an article in a Dutch scientific journal by Dr. Edward Jacobson. He was camping in the forest at the foot of Boekit Kaba on July 10, 1916, when hunters came and told him that they had seen, at a distance of some 20 yards, an *orang pendek* looking for larvae in a rotten stump. The creature was black and agreed with the traditional description of the legendary beast. When it realized that it had been seen it made off, running along the ground on its hind legs. Dr. Jacobson realized that it could not be an orangutan, which would have made its escape from branch to branch of the trees. Near Mount Kerintji, a little farther north, he was able to examine a footprint which his guide said was an *orang pendek*'s, and it was not at all like an orangutan's. It was like a little human foot, but broader and shorter.

A year later, another settler, L. C. Westenenk, produced some more evidence about this creature. One night, around the fire, he heard how there lived in the forests of the Barissan mountains some little men with whom it seemed you had to be on friendly terms. When you went off into the jungle, it was wise always to be provided with tobacco or, failing that, dried moss to use as a substitute, to offer these pygmies if you met them. If you did not leave a handful of tobacco in front of your camp they would torment you all night, making a continual shindy or coming and stealthily pulling branches off your hut. Westenenk had been much amused by these delightful tales, but one day in 1910 the following events altered his opinion:

16. Drawing of the *orang pendek* based on the traditional descriptions.

> A boy . . . employed as an overseer by Mr. van H—— . . . took several coolies into the virgin forest on the Barissan mountains near Loeboek Salasik. Suddenly he saw, some 15 yards away, a large creature, low on its feet, which ran like a man, and was about to cross his path: it was very hairy and it was not an orangutan; but its face was not like an ordinary man's. It silently and gravely gave the men a disagreeable stare and ran calmly away. The coolies ran faster in the opposite direction. The overseer remained where he stood, quite dumbfounded, and when he returned to camp he set down in writing what he had seen. His little note is in my possession.

After citing Dr. Jacobson's account, Westenenk goes on to report what happened to Mr. Oostingh, manager of the coffee plantation at Dataran. At the end of 1917 he lost his way in the virgin forest during an expedition in the eastern foothills of Boekit Kaba. After going round and round in the thick forest for several hours, he saw someone sitting on the ground about 10 yards away who looked as if he were lighting a fire. Oostingh would have been very glad to meet anyone who could tell him how to get home to Dataran, but before he could ask the way he was struck by what he saw:

> I saw that he had short hair, cut short, I thought; and I suddenly realised that his neck was oddly leathery and extremely filthy. "That chap's got a very dirty and wrinkled neck!" I said to myself.
>
> His body was as large as a medium-sized native's and he had thick square shoulders, not sloping at all. The colour was not brown, but looked like black earth, a sort of dusty black, more grey than black.
>
> He clearly noticed my presence. He did not so much as turn his head, but stood up on his feet: he seemed to be quite as tall as I (about 5 feet 9 inches).
>
> Then I saw that it was not a man, and I started back, for I was not armed. The creature calmly took several paces, without the least haste, and then, with his ludicrously long arm, grasped a sapling, which threatened to break under its weight, and quietly sprang into a tree, swinging in great leaps alternately to right and to left.
>
> My chief impression was and still is: "What an enormously large beast!" It was not an orang-utan; I had seen one of these large apes a short time before at Artis [the Amsterdam Zoo].
>
> It was more like a monstrously large siamang, but a siamang has long hair, and there was no doubt that it had short hair. I did not see its face, for, indeed, it never once looked at me.

For lack of a better explanation, Westenenk wondered whether Oostingh had not seen a siamang after all, a great black gibbon which had reached a venerable age.

> Might not the *orang pendek* be the same among the gibbons as the old solitary bulls are among elephants? Lone, ill-tempered great-grandfathers, hated by all the females in their tribe, who have lost most of their hair as a result of mange or age?

This seems to me one of the most reasonable theories that have been put forward. Gorillas, when they grow old, soon become too heavy to go on living in trees. Might not the same be true of gibbons, except that in their case it would have to be for some other reason, such as rheumatic old age?

Unless, of course, the *orang pendek* were a very special sort of gibbon: one that had decided to stick to its two feet, while its cousins still swung acrobatically in the trees; and one which as a result had grown enormously large, or, perhaps, because it had grown so large, no longer felt at home except on the ground. But such a gibbon would be more like a Pithecanthropus—and why not?

Westenenk's article encouraged other settlers to give details of their own experiences. Dr. Jacobson reported the evidence of Mr. Coomans, manager of the State Railway at Padang.

> Once when I was prospecting for minerals in Benkoelen footprints of pygmies were found by one of my European foremen, a man on whose word I could rely, in the valley of the Oeloe Seblat. . . . They were like a child's footprints, but broader. Later the same informant found the same prints near Soungei Klomboek. . . . He noticed the circumstance that along this creature's path the stones had been turned over here and there, as though it was looking for food beneath them.

Dr. Jacobson also spoke of his own experiences. In 1915 at Sioelak Deras several natives told him that they had once or twice seen an *orang pendek*. They all insisted that it was not a siamang; besides, it did not move through the trees but walked on the ground. From Sioelak Deras Dr. Jacobson then went to the edge of the Danau Bento, a huge swamp surrounded by virgin forest. On August 21, 1915, while out in this uninhabited country, where some hunters go to trap deer, his guide Mat Getoep drew his attention to some curious footprints on the bank of a stream (Fig. 17). They could not be confused with the very peculiar trail of the orangutan (Fig. 18). The Sumatran maintained that these prints, which were not more than 5 inches long, had been made by an *orang pendek*.

Captain R. Maier, an official surveyor at Benkoelen, made quite a collection of such footprints. Mr. Lambermon, manager of the coffee plantation at Kaba Wetan, gave him a drawing of two sets of footprints, both found on September 21, 1918, at Roepit and the sides of Boekit Kaba (Fig. 19). Later Mohamad Saleh, a native surveyor in Captain Maier's service, brought him a sketch of a footprint found in the forest at Marga Ambatjung on January 25, 1920. A year later he found another very similar print near Air Roepit (Fig. 20). Another Sumatran surveyor, Raden Kasanredjo, sent Captain Maier a drawing of a double track found near Air Masoedje, made by one animal of the usual size and another much smaller, presumably its young (Fig. 21). There are very clear differences between these prints: while Lambermon's and Raden Kasanredjo's are much alike, Mohamad Saleh's squarer prints seem to belong to a different animal, or at all events to the feet at the other end of the same one, for the hind feet and forefeet of the same creature often produce very different prints. None of these prints seems to be identical with those drawn by Dr. Jacobson.

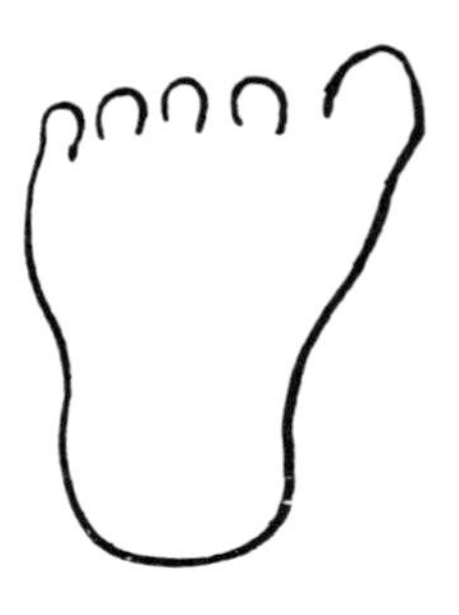

17. Footprint attributed to the *orang pendek,* found on August 21, 1915, by Dr. Jacobson near the Danau Bento swamp.

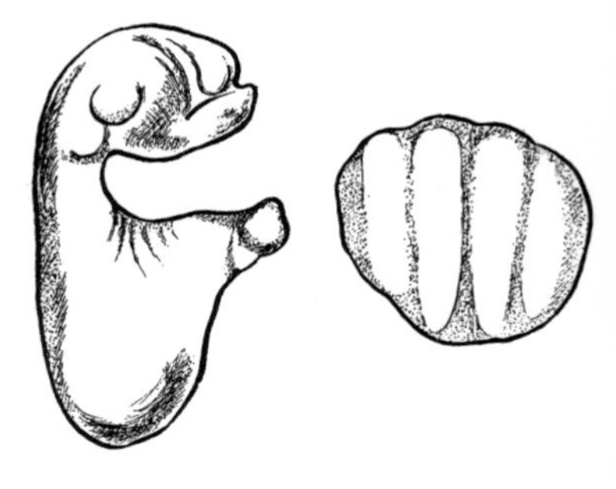

18. Imprints made by the foot (*left*) and second phalanges of the hand (*right*) of a walking orangutan.

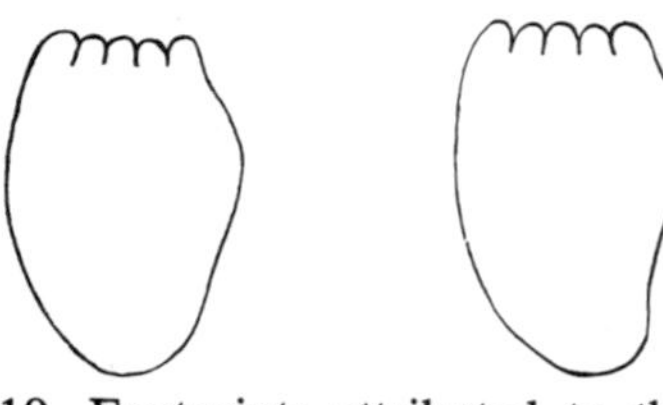

19. Footprints attributed to the *orang pendek* found on September 21, 1918, the left at Roepit and the right on Boekit Kaba, by Lambermon.

20. Footprints attributed to the *orang pendek* found by Mohamad Saleh; (*left*) in the S. Aro forest, January 25, 1920; (*right*) near Air Roepit, January 1921.

21. Footprints attributed to an *orang pendek* and its young, found near Air Masoedje by Raden Kasanredjo.

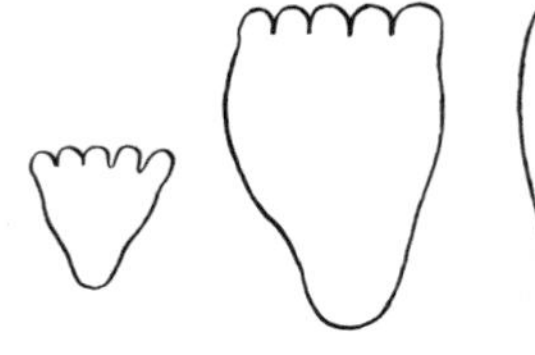

By far the most exact description of the *orang pendek*'s appearance is that due to a Dutch settler called Van Herwaarden. In 1916 he was prospecting for valuable timber in the state of Palembang, and heard tell of a mysterious creature called a *sedapa.* But the accounts were so fantastic and so contradictory that he could not believe that they referred to an unknown "man of the woods."

But at the beginning of 1918 he was exploring the Semangoes district of Moesi Oeloe when he was surprised to see prints of bare feet, very much like a man's, alongside a small river. There were two trails, one large and one small, which he thought must have been made by a mother and her young one. He made a careful sketch of the prints, but it was lost when he accidentally fell into a river. Some days later he met a Mr. Breikers, who had also found similar footprints, and realized that perhaps all the fantastic rumors about the *sedapa* might after all be based on an animal by no means mythical. He therefore began collecting information about it again, and eventually he met three Koeboes who had each seen a *sedapa.* They all said that it was a creature that walked upright and stood between 4 feet 11 inches and 5 feet 3 inches high. It had a hairy body, long hair on its head, and unusually long canine teeth.

Then several years passed without Van Herwaarden hearing anything about the strange beast, until one day he learned that in the forest near the village of Pangkalan Balai, in the Banjoeasin district, a local Malay had recently found the bodies of two dead *sedapas*, a mother and child. He had tried to bring his find back to the village, but the bodies were in such an advanced state of putrefaction that he had to give up this unpleasant task. He died soon afterward; the villagers said it was because he had touched a *sedapa.* And there is no need to bring in magic to see that they may have been right; humping a stinking carcass about through tropical forest is not a healthy occupation, with the jungle scratching at one's limbs and leaving open wounds ready to be infected by one's septic load.

Van Herwaarden writes:

> In October 1923 I was traveling in the same region. . . . For several days from early morning to late afternoon I had been tracking a sounder of wild pig in the cleared part of the island. But, alas, it was to no purpose, although there were countless tracks.
>
> Finally I decided to have one last try and lay in wait, crouching down and well hidden. For an hour nothing happened. Then I hap-

pened by chance to look round to the left and spotted a slight movement in a small tree that stood alone. . . .

My first quick look revealed nothing. But after walking round the tree again, I discovered a dark and hairy creature on a branch, the front of its body pressed tightly against the tree. It looked as if it were trying to make itself inconspicuous and felt that it was about to be discovered.

It must be a *sedapa.* Hunters will understand the excitement that possessed me. At first I merely watched and examined the beast which still clung motionless to the tree. While I kept my gun ready to fire, I tried to attract the *sedapa*'s attention, by calling to it, but it would not budge. What was I to do? I could not get help to capture the beast. And as time was running short I was obliged to tackle it myself. I tried kicking the trunk of the tree, without the least result. I laid my gun on the ground and tried to get nearer the animal. I had hardly climbed 3 or 4 feet into the tree when the body above me began to move. The creature lifted itself a little from the branch and leaned over the side so that I could then see its hair, its forehead, and a pair of eyes which stared at me. Its movements had at first been slow and cautious, but as soon as the *sedapa* saw me the whole situation changed. It became nervous and trembled all over its body. In order to see it better I slid down on to the ground again.

The *sedapa* was also hairy on the front of its body; the color there was a little lighter than on the back. The very dark hair on its head fell to just below the shoulder blades or even almost to the waist. It was fairly thick and very shaggy. The lower part of its face seemed to end in more of a point than a man's; this brown face was almost hairless, whilst its forehead seemed to be high rather than low. Its eyebrows were the same color as its hair and were very bushy. The eyes were frankly moving; they were of the darkest color, very lively, and like human eyes. The nose was broad with fairly large nostrils, but in no way clumsy; it reminded me a little of a Kaffir's. Its lips were quite ordinary, but the width of its mouth was strikingly wide when open. Its canines showed clearly from time to time as its mouth twitched nervously. They seemed fairly large to me, at all events they were more developed than a man's. The incisors were regular. The color of the teeth was yellowish white. Its chin was somewhat receding. For a moment, during a quick movement, I was able to see its right ear which was exactly like a little human ear. Its hands were slightly hairy on the back. Had it been standing, its arms would have reached to a little above its knees; they were therefore long, but its legs seemed to me rather short. I did not see its feet, but I did see some toes which were shaped in a very normal manner. This specimen was of the female sex and about 5 feet high.

There was nothing repulsive or ugly about its face, nor was it at all apelike, although the quick nervous movements of its eyes and mouth were very like those of a monkey in distress. I began to talk in a calm and friendly way to the *sedapa,* as if I were soothing a frightened dog or horse; but it did not make much difference. When I raised my gun to the little female I heard a plaintive "hu-hu," which was at once answered by similar echoes in the forest nearby.

I laid down my gun and climbed into the tree again. I had almost reached the foot of the bough when the *sedapa* ran very fast out along the branch, which bent heavily, hung on to the end and then dropped a good 10 feet to the ground. I slid hastily back to the ground, but before I could reach my gun again, the beast was almost 30 yards away. It went on running and gave a sort of whistle. Many people may think me childish if I say that when I saw its flying hair in the sights I did not pull the trigger. I suddenly felt that I was going to commit murder. I lifted my gun to my shoulder again, but once more my courage failed me. As far as I could see, its feet were broad and short, but that the *sedapa* runs with its heels foremost is quite untrue.

Some most surprising opinions have been expressed about Van Herwaarden's animal. Professor Hill, the expert on the *nittaewo,* thinks that it must have been a gibbon. But quite apart from its size, a gibbon with hair hanging halfway down its back is no ordinary gibbon. Besides a gibbon makes off through the trees, never by running on the ground. And Dr. Dammerman, curator of the Museum at Buitenzorg, remarks that no white man except for Van Herwaarden has ever seen the creature (what about Oostingh?) and adds:

> But this writer is almost too exact in his description of the animal, so that it does not seem impossible that the incident was either based on his imagination, or, that he has written it strongly impressed by the stories about the *orang pendek.*

I leave the reader to judge for himself the credibility of Van Herwaarden's account. Personally I find it refreshingly sober.

The most striking thing about Van Herwaarden's story is the very *human* character of the creature, so that one is rather surprised to find that he keeps calling it a "beast." Its only two features which are not human are its relatively long canine teeth and extremely long arms. Otherwise the description of the creature could apply to a man. The long hair reminds one of the Veddahs and their allied races, but they have bare skin, whereas the *sedapa* was very hairy.

In 1924 the Buitenzorg Museum obtained an actual footprint

supposed to be that of an *orang pendek,* which had been found in Upper Palembang. Molten paraffin wax was poured into the footprint to hold the dry surface earth together, and, when it had set, the thin solidified crust was lifted out of the earth into which the wax had not penetrated.

After examining it with care Dr. Dammerman had little difficulty in showing they belonged to a Malayan sun bear (*Ursus,* or *Helarctos, malayanus*). Moreover, the footprints were found in two very different forms, those of the forefeet and the hind feet, which conclusively proved that the creature was a quadruped. But

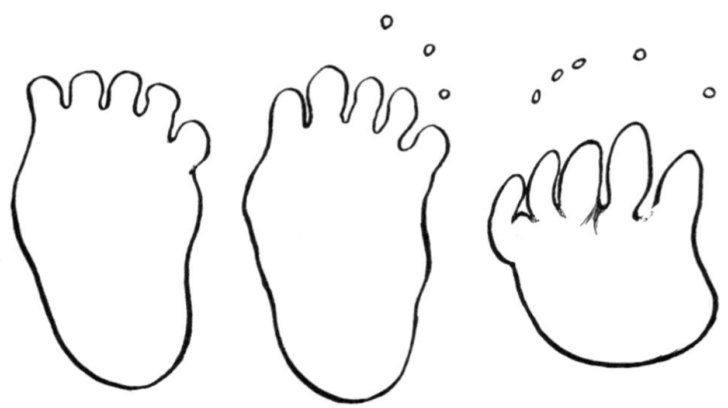

22. Footprint attributed to the *orang pendek* (*left*). It is obviously identical with that of the hind foot of a Malayan sun bear (*center*) whose front footprint is shown on the right.

the mistake was very easily explained, for the perfectly plantigrade hind feet of a bear—"the beast that walks like a man," as Kipling called it—leave prints very like those of human feet. The claws do not always leave a mark, for they are sometimes retracted when the creature is walking.

I have examined lead casts of these prints, and I cannot but agree with Dr. Dammerman's conclusion:

> The peculiar belief that this myterious ape man walks with his heels turned to the front may find its origin in the bear tracks. The Malay bear often turns his feet in, that is to say with his toes turned in and the heel turned out.

Moreover a bear's feet are the "wrong way round" in one sense, for the most conspicuous toe, slightly separated from the others, which seems to correspond to our big toe, is really the little toe, that is to say the outside one.

The Malayan sun bear is, of all the bears, the one which most often stands on its hind feet; the position is habitual. But it never *walks* upright. Can one say that it is responsible for *all* the footprints attributed to the *orang pendek?* Yes, if one maintains that the triangular prints drawn by Dr. Jacobson, Lambermon, and Raden Kasanredjo were made by its forefeet, and the more rectangular prints sketched by Mohamad Saleh by its hind feet. But, while the latter is undoubtedly true, the former is more questionable. The reader can compare for himself the drawing of the print of a bear's forefoot and the various triangular footprints. On the other hand, when a bear is walking it almost always puts its hind feet on the prints left by its forefeet, and so generally leaves a blurred composite trail.

It is clear that some of the reports of the *orang pendek* have been inspired by the footprints of the Malayan bear, and even by this flat-footed creature itself, which stands about 5 feet high on its hind legs. Its very short blackish hair, lighter face, and projecting canine teeth might account for some details reported of the mysterious ape man. An engineer called Nash was disappointed one day when his guide told him to stop because he had spotted an *orang pendek*; he lifted his gun, aimed, fired, and killed a bear. H. de Wals tells of a similar misadventure. What he had taken for an *orang pendek* was an old bear whose neck and back were entirely bald.

23. Malayan sun bear on its hind legs, an attitude it often adopts.

But it is doubtful whether the legend of the *orang pendek* can be based solely on an animal which most Sumatrans know well by the name of *bruan*.

In 1927 a tiger trap in southern Sumatra was found sprung, but the animal which had sprung it escaped without being seen.

> But on the trap [writes Dr. Dammerman] some hair and traces of blood were found; only the discovery was not made at once, but a few days after the animal had escaped. The blood and the hair were examined; it was impossible to obtain any positive results with regard to the hair, but the blood pointed faintly to human origin. . . . it is quite possible that it came from some native who had injured himself while handling the trap.

Why was it "impossible to obtain any positive results with regard to the hair"? The analysis of hair is an advanced science and it should at least have been possible to say what animal could *not* have left the hair. To know, for instance, that it was neither a bear nor an ape would be infinitely precious. Note also the subtlety of the phrase "the blood pointed faintly to human origin." The analysis of the blood must have shown that it was either of human or of animal origin. But that it could "point faintly" to a human origin passes understanding. Does it mean that the creature from which the blood came was *almost* a man? Or that Dr. Dammerman could not bring himself to admit the human characteristics of the *orang pendek* in question? This is what his suggestion of "some native" leads one to believe.

It is clear that the analysts have not been able to identify either the hairs or the blood with those of any known creature.

On May 22, 1932, it looked as if the mystery would at last be solved. The West Sumatran newspaper, the *Deli Courant*, announced that the Rajah of Rokan had surprised two *orang pendek*, one of them a baby. He had fired at them and had brought back the body of the smaller one.

Actually the Rajah had not done the deed himself, though in 1912 he and a friend had seen four *orang pendek* looking for bamboo shoots 15 yards away. They agreed with the usual description and were the size of a child of 12 or 13. The baby had been shot by four natives who had been tempted by the reward offered to anyone who brought back an *orang pendek* dead or alive.

A few days after its first dispatch, the *Deli Courant* published a more detailed report: the beast was 16½ inches high and seemed very human in its anatomy. Its skin was bare, the hair on its head was an even gray. It was sent to the Zoological Museum at Buitenzorg to be examined. On June 9, more details were published in the *Deli Courant* together with a photograph. The creature's arms were rather short; and it could not therefore be an anthropoid ape. On the other hand, it could not be newborn, for its teeth were well developed.

But a few days later the results of Dr. Dammerman's examination of the young "ape man" were published. The whole thing was a colossal hoax. The supposed baby was merely an ordinary lotong (*Trachypithecus*), a sort of langur, which had been shaved all over except on the top of its head. To complete its appearance, its nose had been stretched with a piece of wood, its cheekbones crushed, and its canine teeth filed to a point.

There was nothing new about such hoaxes in Sumatra. Marco Polo had exposed them seven centuries before.

> The country produces a species of monkey, of a tolerable size, and having a countenance resembling that of a man. Those persons who make it their business to catch them, shave off the hair, leaving it only about the chin and those other parts where it naturally grows on the human body. They then dry and preserve them with camphor and other drugs, and having prepared them in such a mode that they have exactly the appearance of little men, they put them into wooden boxes, and sell them to trading people, who carry them to all parts of the world.

Annoyed at being taken in by the same trick that had been played on innocent travelers in the Middle Ages, the world of science has ever since remained obstinately silent about the *orang pendek*, and shelved it with the unicorn and the phoenix.

But is this logical? The fact that a forger of genius painted "Vermeers" which took in experts of the highest repute does not mean that the great Dutch painter never existed or that he did not paint his own pictures.

CHAPTER **6**

THE NOT SO ABOMINABLE SNOWMAN

There was exactly the very Print of a Foot, Toes, Heel and every Part of a Foot; how it came thither, I knew not, nor could in the least imagine . . . nor is it possible to describe how many various Shapes affrighted imagination represented things to me in . . .
DANIEL DEFOE, *Robinson Crusoe.*

AFTER HIS SIXTH ATTEMPT on Everest, Eric Shipton was exploring the neighboring Gauri Sankar range with Michael Ward and the Sherpa Sen Tensing. At four o'clock in the afternoon of November 8, 1951, they found a very clear trail of enormous human-looking feet in the powdery snow on the southwestern slopes of Menlungtse. They followed this strange trail for about a mile until they lost it in a moraine of ice. Being unable to follow the mysterious creature any farther, they took photographs of its footprints.

Roughly oval in shape, they seemed to have been made by human feet—but feet more than a foot long. Shipton remarks that they were "slightly longer and a good deal broader than those made by our large mountain boots." A man on this scale would stand about 8 feet high. The big toe was clearly visible, slightly separated from the rest, but there seemed to be only three other toes. Of course it was possible that two toes might be held so close together that they left only a single print in the snow. "Where the tracks crossed a crevasse," Shipton goes on, "one could see quite clearly where the creature had jumped and used its toes to secure purchase on the snow on the other side."

Ever since 1899, strange rumors had been reaching the West about the giants that lived in the icy heights of the Himalayas. In that year appeared Major L. A. Waddell's *Among the Himalayas,* in which he tells how in 1889 he found large footprints in the northeast of Sikkim: "These were alleged to be the trail of the hairy wild men who are believed to live amongst the eternal snows."

During the first attempt to climb the North Face of Everest in 1921 the rumors became more detailed. On the way from Kharta to the pass at Lhapka-la, Colonel Howard-Bury and his companions saw dark spots moving over the snow in the far distance. Then, on September 22, 1921, when they reached the place, about 23,000 feet up, where they had seen them, they found enormous footprints. The leader of the expedition attributed them to a large stray gray wolf, but the Tibetan porters had no doubt that these were footprints of the *metoh kangmi* or "abominable snowman."

From the native reports, which agree in the main, one can compile the following description of the monster. The snowman is a huge creature, half man, half beast; it lives in caves high and inaccessible in the mountains. The skin of its face is white; the body is covered with a thick coat of dark hair. Its arms, like those of the anthropoid apes, reach down to its knees, but its face looks rather more human. Its thick legs are bowed; its toes turn inward—some even say they turn backward. It is very muscular and can uproot trees and lift up boulders of remarkable size.

These legends are found for thousands of miles all over the Himalayan range, from the Karakoram to northern Burma, in Tibet, Nepal, Sikkim, Bhutan, and Assam, and the creatures have many different names in different countries.

Most naturalists dismissed these legends. But uneducated Tibetans or Nepalese were not the only people to spread this story.

In an interview in *The Times* of November 2, 1921, an Englishman called William Knight told of a strange encounter near Gangtok on the way back from Tibet shortly before the last Tibetan war:

> I stopped to breathe my horse on an open clearing, and loosened the girths, and watched the sun, which was just about setting. While I was musing, I heard a slight sound, and looking round, I saw some 15 or 20 paces away, a figure which I now suppose must have been one of the hairy men that the Everest Expedition talk about, and the Tibetans, according to them, call the Abominable Snowman.
>
> Speaking to the best of my recollection, he was a little under 6 ft high, almost stark naked in that bitter cold—it was the month of November. He was a kind of pale yellow all over, about the colour of a Chinaman, a shock of matted hair on his head, little hair on his face, highly splayed feet, and large, formidable hands. His muscular development in the arms, thighs, legs and chest was terrific. He had in his hand what seemed to be some form of primitive bow.

In 1925 a Fellow of the Royal Geographical Society called N. A. Tombazi reported a no less startling story in his *Account of a*

24. Reconstruction of the abominable snowman based on the most detailed evidence available.

Photographic Expedition to the Southern Glaciers of Kangchenjunga in the Sikkim Himalaya. About 9 miles from the Zemu glacier, at an altitude of some 15,000 feet, he noticed his porters waving and pointing at an object lower down.

The intense glare and brightness of the snow prevented me from seeing anything for the first few seconds; but I soon spotted the "object" referred to, about two to three hundred yards away down the valley to the East of our camp. Unquestionably, the figure in outline was exactly like a human being, walking upright and stopping occasionally to uproot or pull at some dwarf rhododendron bushes. It showed up dark against the snow and, as far as I could make out, wore no clothes. Within the next minute or so it had moved into some thick scrub and was lost to view.

Such a fleeting glimpse, unfortunately, did not allow me to set the telephoto-camera, or even to fix the object carefully with the binoculars; but a couple of hours later, during the descent, I purposely made a detour so as to pass the place where the "man" or "beast" had been seen. I examined the footprints which were clearly visible on the surface of the snow. They were similar in shape to those of a man, but only six to seven inches long by four inches wide at the broadest part of the foot. . . .

When I asked the opinion of the Sirdar and the coolies they naturally trotted out fantastic legends of "Kangchenjunga-demons." Without in the least believing in these delicious fairy-tales myself, notwithstanding the plausible yarns told by the natives, and the references I have come across in many books, I am still at a loss to express any definite opinion on the subject.

Tombazi nevertheless suggests his own theory, somewhat tentatively, since it does not agree with the shape of the footprints, which were too broad in proportion to their length to be a man's:

I conjecture then that this "wild man" may be either a solitary or else a member of an isolated community of pious Buddhist ascetics, who have renounced the world and sought their God in the utter desolation of some high place, as yet undesecrated by the world. However, perhaps, I had better leave these conclusions to ethnological and other experts.

The explanation often put forward that the snowman is a *sadhu,* one of those Hindu hermits who do in fact live at altitudes up to 15,000 feet in the Himalayas, is superficially attractive. It might possibly account for such reports as Knight's and Tombazi's, but it agrees neither with the beast's alleged appearance nor with the size and shape of the tracks it leaves in the snow.

Inevitably most of the descriptions of the snowman come from very simple people: peasants and porters, who are terrified of it. Frank Smythe tells how during the 1930 Kangchenjunga expedition the noise made by a yak bursting in unexpectedly set his porters in a real panic; they thought they were being attacked by snowmen. When he learned the actual cause of the commotion, one of the Sherpas called Nemu told Smythe that he had several times seen "*bad manshi*," as he called them in pidgin English, with his own eyes. He said that they were huge white men covered with thick fur. Miss MacDonald of Kalimpong, the daughter of David MacDonald, the famous anthropologist, also told Smythe how, when she was passing through a defile at a great height on a journey to Tibet, she heard a terrifying roar, unlike any animal's cry that she had ever heard. Her porters were panic-stricken; they dropped their loads and left her there alone. According to the numerous witnesses who have described the snowman's cry—some even trying to imitate it—it is a "loud yelping," often compared to the sad sound of "the mewing of a sea gull," but much louder.

Western explorers had no way of verifying the snowman's existence except by studying the strange tracks of bare feet in the high snows. During the summer of 1931 Wing-Commander E. B. Beauman saw them on a glacier some 14,000 feet up near the source of the Ganges. In 1936 Eric Shipton also saw them 16,000 feet up on his return from Everest. "They resembled a young elephant's tracks except that the length of the stride suggested a biped."

A little later the well-known ethnographer and botanist Ronald Kaulback also met them looking "exactly as though they had been made by bare-footed men" some 16,000 feet up in the southeast of Tibet, on the main route between the valleys of the Ge-chu and the Upper Salween. As there were no bears in the region as far as he knew, he thought they must have been made by a snow leopard (*Panthera,* or *Uncia, uncia*). The porters, of course, at once spoke of "Mountain Men." And one of them who had happened to see one from fairly close described it as like a man with white skin, naked, with long hair on its head, shoulders, and arms. Kaulback thought that the lack of food at this altitude was enough to refute the legend, but this applied equally well to the snow leopard.

In 1937, 20,000 feet up in the Bhyundhar Valley, Frank Smythe found footprints in the snow, and followed the trail up to the entrance of a cave. He gives a strange description of the prints.

> On the level the footmarks averaged 12 to 13 in. in length and 6 in. in breadth, but uphill they averaged only 8 in. in length. The stride was some 1½ to 2 ft. on the level, but considerably less uphill, and the footmarks were turned outward at about the same angle as a man's.* There were well-defined imprints of five toes, 1½ inches to 1¾ inches long and ¾ of an inch broad, unlike human toes, arranged symmetrically. Lastly, there was what appeared to be the impression of a heel with two curious toelike impressions on either side.

* They were not, as I shall explain, but they looked as if they were.

Footprints of this kind gave rise to the belief sometimes held by the Sherpas that the *yeti* has extra toes and walks with its feet back to front. This is yet another version of the legend current in Sumatra about the *orang pendek*; it was believed since time immemorial, for even in Megasthenes's *Indica* of the fourth century B.C. we read, "In the mountains called Nulo there are men whose feet point backward and have eight toes on the ends."

Smythe's photographs of the trail showed that it was indisputably a bear's. The marks of the extra toes were really those of the side toes of the hind feet, for when a bear is walking it usually puts its hind feet down in the footprints of its forefeet. Moreover it turns its feet inward, so that from the position of the prints alone the trail looks as if it is going in the opposite direction.* Then the toes are seen to be on the wrong end of the foot, and so the legend of the men with their feet back to front arose.

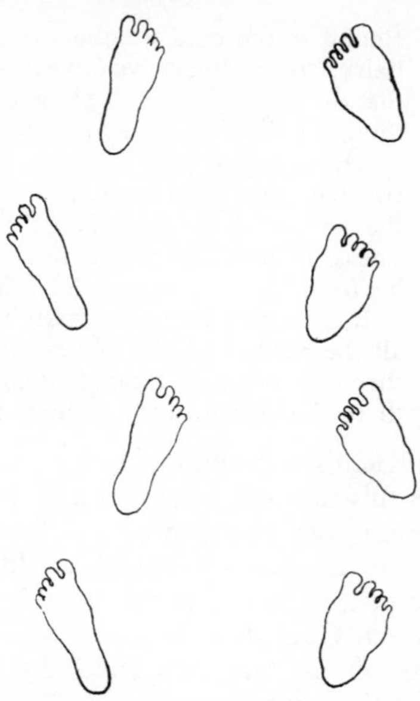

25. Comparison between a human trail (*left*) and a bear's (*right*). Actually a bear's prints are never so clear, since the prints of the hind feet are rarely exactly superimposed on those of the forefeet. At a superficial glance the bear's trail may look as if it is walking *down* the page, with its toes turned *outward*.

To many zoologists it seemed that the affair of the abominable snowman could now be shelved. But this view was premature: true snowman's footprints never show the same characteristics as a bear's trail. All the same, the idea that its feet were turned backward seemed to be firmly rooted in local tradition. A curious letter written in 1915 by J. R. O. Gent, a forestry officer in the Darjeeling Division, should be included in the snowman's dossier.

> I have discovered the existence of another animal but cannot make out what it is, a big monkey or ape perhaps—if there were any apes in India. It is a beast of very high elevations and only goes down to

* A bear's feet are also "back to front" in another sense (see page 72).

Phalut in the cold weather. It is covered with longish hair, face also hairy, the ordinary yellowish-brown colour of the Bengal monkey. Stands about 4 feet high and goes about on the ground chiefly, though I think it can also climb.

The peculiar feature is that its tracks are about eighteen inches or two feet long and toes point in the opposite direction to that in which the animal is moving. The breadth of the track is about 6 inches. I take it he walks on his knees and shins instead of on the sole of his foot. He is known as the jungli admi or sogpa. . . .

It is a thing that practically no Englishman has ever heard of, but all the natives of the higher villages know about it. All I can say is that it is *not the Nepal Langur,* but I've impressed upon people up there that I want information the next time one is about.

Gent's explanation is ingenious, but no known monkey moves in this unusual way; and if it did it would not leave separate footprints, but two continuous furrows.

As a biped 4 feet high could hardly leave such enormous footprints, it seems that they could not have been made by the monkey which Gent describes, and which is probably the Himalayan langur, unless they were made by its four feet together. I shall return to this point later.

Subsequently, many other Himalayan explorers discovered strange footprints in the snow at unusually high altitudes. Among them was a correspondent of *The Times* who signed himself "Balu." In 1937 he was surveying in the Karakoram, in the north of the Himalayan range, when he was brought up short by a perfect row of large footprints, more or less round and about a foot in diameter. They were about 9 inches deep in the snow and 18 inches apart.

In the same year, John Hunt discovered footprints in the Zemu Gap. Then in 1938, H. W. Tilman found them in the same place during a new attempt on Everest.

It was on one of the glaciers of the Menlung basin, at a height of about 19,000 feet, that, late one afternoon, we came across those curious footprints in the snow the report of which has caused a certain amount of public interest in this country. We did not follow them further than was convenient, a mile or so, for we were carrying heavy loads at the time, and besides we had reached a particularly interesting stage in the exploration of the basin. I have in the past found many sets of these curious footprints and have tried to follow them, but have always lost them on the moraine or rocks at the side of the glacier. These particular ones seemed to be very fresh, probably not more than 24 hours old. When Murray and Bourdillon followed us a few days later the tracks had been almost obliterated by melting. Sen Tensing, who had no doubt whatever that the creatures (for there had been at least two) that had made the tracks were "Yetis" or wild men, told me that two years before, he and a number of other Sherpas had seen one of them at a distance of about 25 yards at Thyangbochi. He described it as half man and half beast, standing about five feet six inches, with a tall pointed head, its body covered with reddish brown hair, but with a hairless face. When we reached Katmandu at the end of November, I had him cross-examined in Nepali (I conversed with him in Hindustani). He left no doubt as to his sincerity. Whatever it was that he had seen, he was convinced that it was neither a bear nor a monkey, with both of which animals he was, of course, very familiar.

At a reception given at the British Embassy in Katmandu, Sen Tensing confirmed this description, adding that "it moved mostly in an upright stance but when in a hurry dropped on all fours."

The snowman became news: in 1952 Prince Peter of Greece, who was engaged in anthropological research in India and Tibet, wrote a long letter to the Indian newspaper, the *Statesman,* for he too had recently picked up some fairly detailed information about the mysterious creature.

For some time one of these snowmen—which were actually large monkeys—had been in the habit of coming in the night and drinking from a cistern at the mouth of the Jalap valley in Sikkim. Alarmed at the mere idea of sharing their drinking water with such an unwelcome visitor, the villagers prepared for it a bucket full of fermented liquor. The brute came, drank as usual, and eventually collapsed, dead drunk. The men found it in the morning and lashed it firmly to a pole. When it sobered up it recovered not only its senses, but also all its strength; it burst its bonds and escaped.*

When Dr. Wyss-Dunant's Swiss expedition made its assault on Everest in 1952 it also found footprints exactly like those that Shipton had photographed the year before. On April 18, René Dittert, André Roch, and the famous Sherpa Tenzing Norkey set off on a reconnaissance along a glacier. There was a pea-soup fog. When the three men came back they found that at an altitude of 19,000 feet their own trails crossed those of a group of *yetis,* which had perhaps been shadowing them in the fog.

* Professor René von Nebesky-Wojkowitz tells another version of this story. It was said to have happened at the Natu Pass, an adjacent pass to Jalap-la. Otherwise the story is identical and may be taken as confirmation of Prince Peter's version.

On the clearest footprints one could see a separate big toe and four other toes. Dr. Wyss-Dunant declared that they had been made not by a biped but by a quadruped, probably related to a bear and weighing between 168 and 217 pounds.

In 1953, when the successful Everest expedition was at Thyangboche, Sir John Hunt made inquiries about the *yeti* at the monastery. He learned that the monks occasionally saw snowmen on the heights above their settlement, and did not think there was anything mysterious about them. The last time they had seen one, in November 1949, it came out of a large clump of rhododendrons and played for some time in the snow no more than 200 yards away. "It was," Sir John reports, "a largish animal, five feet or more in height, covered with greyish-brown hair. It went mainly upright and occasionally dropped on all fours: it was also seen to scratch itself monkey fashion." This report merely confirms the main details of Sen Tensing's; for he was present on this occasion with several other Sherpas.

Like all animals, the *yeti* is sacred, and its remains are worshiped as relics in lamaseries. This was how a Tibetan lama called Chemed Rigdzin Dorje Lopu maintains he was able to examine the mummified bodies of two of these creatures, one in the monastery at Riwoche in the province of Kham, the other in the monastery at Sakya on the road from Katmandu to Shigatse. They were enormous monkeys about 8 feet high. They had thick flat skulls and their bodies were covered with dark brown hair about 1 inch to 1½ inches long. Their tails were extremely short.

What did he mean by "thick flat skulls"? Sen Tensing said it had "a tall pointed head," and the famous Sirdar Tenzing Norkey confirmed this description when he told Sir John Hunt how his own father had once met a *yeti* "at the yak-herds' village of Macherma" at the mouth of the Dudh Kosi.

The elder Tenzing was driving his herd to pasture in the valley when he saw a sort of little hairy man who was rushing down the mountainside in leaps and bounds. Tenzing was terrified and led his yaks to a stone hut, but the *yeti* was furious and leaped on the roof and started tearing off the shingles, which were merely held down with stones. The shepherd was forced to light a fire, which gave off acrid smoke and eventually drove away the *yeti,* but not until it had let off its fury in a typically monkey fashion by dashing chattering round the hut, tearing up small shrubs and hunks of rock.

According to the elder Tenzing, the animal walked upright like a man, was about 5 feet high, and its body was covered with reddish-brown fur. It had a large ape's features, "but the mouth was especially wide, showing prominent teeth." Its skull was high and conical in shape, and covered with hair so long that it fell down in front of its eyes.

It therefore seems as if Chemed Rigdzin meant by "thick, flat skulls" that the sides were particularly flat, making them pointed and conical in shape.

This was how the snowman's dossier stood at the end of 1953. On the whole it was treated with utter disbelief. But the footprints in the snow could not be imaginary. How did science explain them? Colonel Howard-Bury attributed them to a gray wolf and Ronald Kaulback, though with some reservations, to a snow leopard. Neither hypothesis accords with the shape of the footprints. But in 1937 better explanations had been put forward in *The Times*.

Guy Dollman pointed out that in the Himalayas there were two species of monkeys of reasonable size. One was the Himalayan

26. Roxellana's snub-nosed langur.

langur (*Semnopithecus schistaceus*): it had been recorded in Tibet up to an altitude of 13,000 feet. It has a tougher constitution and longer and fleecier fur than the commoner hanuman. The other was

27. Himalayan langur.

Roxellana's snub-nosed langur (*Rhinopithecus roxellanae*) which the Chinese called the snow monkey. It was a primate of much heavier build than the langur, with thick fur and a comically snub nose which made its face look very human. These monkeys were very rare and then still unknown in captivity.

The size of these creatures does not agree with the traditional description of the snowman, for the largest hardly ever exceed 4 feet 6 inches in height when standing on their hind legs. Moreover, the snub-nosed monkey is found in country to the northeast of the Himalayan range, eastern Tibet and western China, whereas it is on the southwestern slopes that we seem to hear most about the snowman.

All the same, most zoologists favored the monkey theory when Eric Shipton brought back his famous photographs. They were supported by no less an authority than Dr. T. C. S. Morrison-Scott, and went so far as to specify the race of langurs the pseudo-monster belonged to. It was a *Presbytis* [alias *Semnopithecus*] *entellus achilles,* they said, a langur which may stand as much as 5 feet high. It was a graceful quadrumane, mainly fawn in color with a black face and silvery head and neck. Its description, these zoologists maintained, agreed fairly well with that given by Sen Tensing, Shipton's guide.

Against this theory it was shrewdly objected that the footprints of a langur with long toes were not in the least like those in Shipton's photographs made by a large solid foot with short toes, and in any case an animal with feet no more than 9 inches long could not leave footprints 13 inches long. This latter objection was ingeniously evaded by an argument which a newspaper put thus:

> The colossal dimensions of the footprints are no doubt to be explained by the fact that they are made in the night in fresh snow and in the day-time they melt in the snow and increase in size.

Unfortunately, a few lines above, the same newspaper rashly observed that "the footprints are so far apart that it is impossible that they could have been made by a human being." A little thought should have made it obvious that if a small animal's footprints were enlarged by the snow melting the length of its stride would appear to decrease, the prints might even merge.

Oliver Jones, Curator of Mammals at the London Zoo, put forward a very ingenious theory that the footprints Shipton photographed had been made by *all four feet at once*. As it bounded across the snow the langur first put down its two hands side by side and then its hind feet, also held close together, immediately behind them. This would produce a string of large bilobate footprints.

This explanation seemed to be confirmed by W. W. Wood, who wrote to *Country Life* about a trip in the mountains at Liddarwat near Srinagar that he made in 1944 with Major Kirkland and Captain John B. Maggs.

> We were somewhere near the timber line—about 13,000 feet—when we saw a large animal bounding towards us down the snow-covered khud on the opposite side of the river. Its gait appeared to me to be that of a monkey in a hurry, with all four paws off the ground together. Maggs' recollection is that its rear legs were longer than the forelegs and its running not very different from that of a rabbit . . .
>
> It was tawny in colour, with a fringe round its face, was about the size of a man and had a long tail with a tuft on the end, like a lion. . . . The Kashmiris said that it was *bandar* (monkey).
>
> This strange creature was certainly neither bear nor langur. Can it, wandering alone at these altitudes, have been abominable snowman?

To which the editor replied:

> Despite our correspondent's assertion to the contrary, the creature was almost certainly a langur, for he gives a good description of *Semnopithecus entellus ajax* . . . Admittedly this langur measures only about 30 inches head and body, but size is notoriously hard to judge in the mountains.*

All the same Wood's description of the monkey's strange progress in leaps and bounds corroborates Oliver Jones's theory. The snag is that some of Shipton's prints are too clear to have been made by four feet merged into one. You can distinctly see the general outline of the foot and the details of most of the toes. Moreover if you look clearly at the photograph of the trail you can see that the footprints alternate to left and right. This is proof that the animal is *walking,* for why the devil should an animal jump alternately to left and to right with absolute regularity? And a monkey like a langur would have left marks in the snow with its tail.

The party who maintained the *yeti* was a bear were in a stronger position, especially after Frank Smythe's photograph. According to

* This is very true, and one may as well underestimate the size as exaggerate it. Thus the *yeti* could easily be larger than the 5 feet 10 inches to 6 feet 6 inches which some witnesses have reported. The huge size of its footprints would seem to show that it is.

their champion, Reginald I. Pocock, the creature responsible for the giant footprints could only be the red bear (*Ursus arctos isabellinus*), a local Tibetan species of the European brown bear. This bear, which rarely exceeds 6 feet 6 inches in height, could easily leave footprints 13 inches long, and what is more they would be astonishingly like a man's.

This argument was firmly supported, among others by G. S. Cansdale, Superintendent of the London Zoo, and there was much in its favor. What, one might reasonably ask, could an entirely vegetarian monkey—a specialized leaf-eater like the langur—find to eat at an altitude of 19,000 to 23,000 feet? For these monkeys have never been reported above 13,000 feet—that is, above the tree line—and this height is very exceptional.

For bears the problem of food does not arise the whole year round. During the bad season they leave off feeding and become lethargic and sluggish. Thus there is no theoretical reason why some species of bear should not spend the cold months of the year at high altitudes where it would be absolutely safe from attack.* And it is at the beginning and at the end of the hibernation season that footprints are usually found at unusual heights. The snowman generally leaves its tracks in autumn snows.

So far the bear theory seems very satisfactory. Indeed I favored it at first myself after carefully comparing Shipton's photographs with various bears' footprints. Some so-called experts maintained that if the snowman were a bear, one would have seen signs of its claws. This is not so at all. The plantigrades stand mainly on the fleshy part of the sole of the foot and on the toes, and their claws do not always touch the ground. If it is indeed a bear one can even specify which of the bear's feet made the print in Shipton's clearest photograph. First of all it is plainly a *hind foot,* for the forefeet leave a much rounder print (see Fig. 22, page 72). Second, if it was really made by a bear it was by the *right* hind foot—for as I have already remarked what looks like a bear's big toe is actually its little toe. Therefore an expert would only have to glance at the trail to see whether the "big toe" was on the outside or the inside to know whether it was that of a bear or of some kind of primate.

It is a pity, therefore, that Shipton did not take a close-up of several successive footprints instead of just publishing a close-up of a single print and some general views of the trail as a whole.

* In Europe, it is true, the contrary happens: bears leave the heights in autumn and come down into the valleys to hibernate.

All the same, when I examined them carefully they told me that it was wrong to think the creature was a bear on the sole evidence of a single footprint, however clear. For the position of the footprints is very clear: an almost straight track with each print a little off-center, alternately to left and to right. The first explanation that springs to mind of such a trail is that it was made by a biped. Thus:

left foot   left foot
        right foot   right foot

But is it impossible for a quadruped to make a trail like this? Certainly such a close and regular trail could not be caused by a *gallop*. So there is no need to worry about the very peculiar way in which a quadruped puts its feet down when moving fast. The trail that a quadruped makes when walking or trotting can be considered as if it were made by two bipeds, one behind the other, moving at the same speed. It makes no difference whether it ambles or walks in the usual way, the feet may be put down at different times, but not in different places.

The snowman's trail could not have been made by a quadruped with front feet a different distance apart than its hind feet or they would be out of line. A mouse's trail shows this effect

28. Mouse's trail.

very well: there are four lines of footprints. But the snowman's trail is in only two lines.

This leaves three possible kinds of quadruped's trail:

1. The animal might put its hind feet down exactly in between the footprints of its front feet, thus:

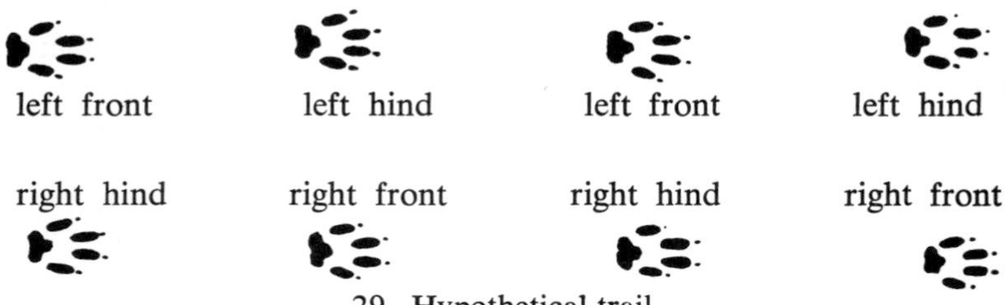

29. Hypothetical trail.

2. Most animals put their hind feet down nearer to their front feet, thus:

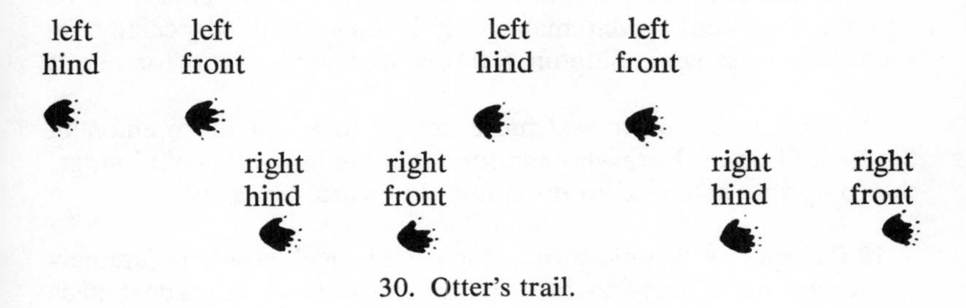

30. Otter's trail.

3. Finally the animal might put its hind feet directly on top of the prints of its front feet, thus:

(left hind and front) (left hind and front) (left hind and front)
(right hind and front) (right hind and front)

Only the third of these kinds is arranged like the snowman's trail.

Every single mark is made by two footprints superimposed, and therefore the track is made up entirely of blurred prints. Now Shipton's footprints are clearly outlined in the snow and their shape is very regular. They are not, as some have maintained, half obliterated and blurred by the wind. Shipton has stressed that they were fresh. Therefore unless the footprint so carefully photographed was an exceptional one—for instance when the animal was leaping across a crevasse—the track could only have been made by a biped. "Balu" too insisted that in the track discovered in 1937 there was no "sign of overlap, as would be the case with a four-

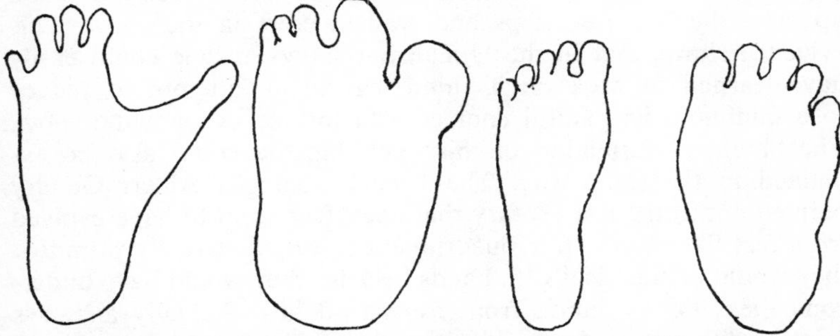

31. Outline of the left footprint of (*from left to right*) gorilla, snowman, man, and bear.

footed beast." So if the animal was a bear, it was a bear that walked on its hind legs. It is indeed true that bears occasionally adopt an upright gait when they are inquisitive or threatening; but I do not think they would ever do so for several hundred yards or that they could even manage it if they tried—especially on such sloping ground. Shipton followed the *yeti*'s track for about a mile.

It seems much wiser and more honest to adopt the opinion of Sir John Graham Kerr who admitted that, as far as he could judge, the footprints belonged to no animal known to zoologists.

If the species is unknown, which zoological group of animals is the *yeti* most likely to belong to? The only plantigrade bipeds are man and those tree-dwellers the gibbon and the indris—all primates. Could the *yeti* be a large anthropoid ape as yet unknown?

There are two points against this hypothesis. First, that all the anthropoids have an opposable big toe—like a thumb—clearly separated from the other toes, indeed almost at right angles to them when the foot is put on the ground. Second, the anthropoids are usually quadrupeds. They are not built to walk on their hind legs. They always tend to fall over forward, and only crutches or balancing rods can prevent them from doing so. Their arms fulfill this purpose.

Nevertheless it is not impossible for anthropoid apes to walk on their hind legs, even if they find it hard going. With patience it is possible to train a chimpanzee to walk some distance on its hind legs. Certain climates may even naturally encourage such a gait. Professor Sydney W. Britton describes how snow had fallen, and he was curious to see how a chimpanzee would react to a phenomenon it had never seen before. To his surprise it stood up after the first few steps and walked erect in the snow. This evidence shows that in the Himalayan snows an ape could easily have learned to walk on its hind legs in an attempt to reduce to a minimum its painful contact with the ice beneath the snow. The "human" formation of this ape's big toe could also be explained in the same way. The French zoologist Albert Gaudry pointed out in the last century that apes' feet seem to have evolved from feet like man's. If a quadrumane's feet were really primitive they would be identical with hands because they would have undergone the same evolution from fish's oval fins. Actually an ape's foot is still a true foot with a well-developed heel, lengthened

32. Orangutan walking in an upright, but still quadruped, position.

and curved toes, and a big toe which has gradually separated from the rest and become opposable.

A primate that, like man, had not become specialized to living in trees would, like him, have retained truly plantigrade feet. It is easy to see how if a line of apes evolved in increasing size they would very soon have to give up living in trees. The mountains would be an ideal habitat, and there the snow would encourage them to become bipeds. Thus, in theory at least, a race of giant apes, with primitive plantigrade feet and a tendency to stand on their hind legs, could have arisen in the mountain snows.*

So, in the end, had we not better humbly accept the evidence of those who describe the snowman as a giant ape, a sort of hairy ogre? Most people may reply that giants have never existed. But this, as we shall see, is not true.

In 1934 Ralph von Koenigswald, a young Dutch geologist and paleontologist, who later contributed so much to our knowledge of the Java ape man, was wandering about the streets of Hong Kong looking for curiosities. He went into an old-fashioned Chinese chemist's shop. He had come to China to study its fauna, and among the chemist's junk one sometimes came upon stuffed specimens of rare animals, dried insects, shells, or even pieces of fossils. On the counter he noticed a jar full of teeth of all sorts and picked up a handful. It was child's play for him to recognize what animal they belonged to, for teeth, especially those of mammals, are like identity cards to an expert. Suddenly he stopped with a shiver, so astounded, as he later recalled, that his hair actually stood on end. He held in his hand a tooth that looked to be human —he could tell that it was a third lower molar—but a tooth far larger than any man or ape had ever possessed. Its volume was five or six times greater than the corresponding tooth of a man. It was a giant's molar.

"Where did you get that?" he asked the chemist in a faint voice.

The Chinese did not know. It had been in that batch of teeth for a very long time. No doubt his father or his grandfather or even some more distant ancestor had got it from some peasant. They often found "dragon's teeth" in the fields.

Von Koenigswald set about scouring every shop in the district. Two years later, across the water in Canton, he found another tooth similar to the first, this time an upper molar. In 1939, in just the same way, he finally added to his collection another third lower molar, much better preserved than the first, for it still had the root as well as a perfectly intact crown. All these teeth, while undoubtedly belonging to a primate, were twice as big as those of an adult male gorilla. If the rest of the giant were in proportion it must have stood between 11 and 13 feet high.

* This argument can be checked by applying it to a hypothetical carnivore instead of a hypothetical primate, and reaching the conclusion that a race of giant carnivores, with primitive plantigrade feet and a tendency to stand on their hind legs, could have arisen in the mountain snows. They have. They are bears.

33. First upper molar of Gigantopithecus (*right*) compared with the corresponding tooth in man.

Was it man or ape? It is hard to say; for the teeth of men and the larger anthropoids are much alike. A considerable difference in size between creatures belonging to the same zoological order usually involves numerous variations in the details of their anatomy. The distinction between ape and man here seems premature. It is therefore more prudent to say no more than that the creature which Von Koenigswald called *Gigantopithecus* (or "giant ape") was a giant ape man.*

Von Koenigswald's discoveries about giants did not end there. In Java in 1941 he dug up a fragment of an enormous jawbone

34. Fragment of the jaw of Meganthropus (*left*) compared with the corresponding piece of a human jaw.

in which there were still three teeth. They were even more human in appearance than those of the Gigantopithecus, but were only

* A jawbone belonging to the *Gigantopithecus* was recently found in a mountain cave in Kwangsi Province, South China. Dr. Pei Wen-Chung reports that it is definitely of an ape, not a man. It is between 400,000 and 600,000 years old (Middle Pleistocene period). Wear on the teeth shows "the animal had a mixed diet of meat and vegetables, quite different from that of modern apes which live on fruit." Dr. Pei adds: ". . . this anthropoid was closer to man than any other ape yet discovered. It is estimated to have had a height of some 12 feet."

three quarters of the size. The creature to which they belonged must, by this reckoning, have stood between 8 and 10 feet high. It was duly christened *Meganthropus palaeojavanicus,* the Great Man of Old Java.

Then, in April 1948 J. T. Robinson, assistant to Dr. Robert Broom of the Transvaal Museum, unearthed in a cave at Swartkrans, in South Africa, the larger part of a huge jawbone, still containing three premolars and four molars. Nearby were an upper wisdom tooth, two incisors, and an upper canine tooth. The jaw was very slightly smaller than that of the Meganthropus. The premolars were one and a half times as big as a man's, but strange to say the incisors and canines were of the same size as the corresponding human teeth. Dr. Broom called this African cousin of the Meganthropus by the name of *Paranthropus crassidens*. It is the best known of all giant primates. In 1949 and 1950 Broom and Robinson unearthed a good dozen fragments of its bones, including two almost complete skulls and a pelvis which showed that the creature stood more or less erect. The Paranthropus's face did not have such a prominent muzzle as the anthropoid ape's. Its face was broad, fairly flat, and apparently almost noseless. By far its most striking feature were its huge jaws, which contrasted oddly with its relatively small cranium. It is not therefore surprising to find that this huge-jawed creature had a crest along the top of its skull like that to be seen on old male gorillas, for these crests are closely linked to the muscular strength of the jaw, the terminal tendons of which are attached to the parietal bones near the top of the skull.

From the dimensions of the jawbone and the remaining parts of the skull, Broom estimates that the cranial capacity of *Paranthropus crassidens* is more than 900 cubic centimeters.

> Quite possibly the brain of a large male Swartkrans ape-man may be over 1,000 c.c., and thus as large as the brains of Leibniz or Anatole France.

He thus concluded that he had at last found the true "Missing Link." But, one may object, it signifies nothing that a giant should have a larger brain than man; whales and elephants have much larger brains but their intelligence certainly is not in proportion. No doubt it was in order to forestall this objection that Dr. Broom later maintained that his Paranthropus was no bigger than a man or at the very most a gorilla, and differed only in having abnormally developed jaws.

But, as Weidenreich pointed out,

> In most cases, and especially in the Primates, large teeth necessitate large jaws, and large jaws necessitate a large body, as we know from fossil giant lemurs of Madagascar.

Therefore until there is proof to the contrary it seems more reasonable to suppose that the Swartkrans Paranthropus is a large ape man with powerful jaws and a relatively small brain. As to the giants of Java and China, there is no proof that their jaws were disproportionately large since we still know nothing about the formation of their skulls. But even if the Gigantopithecus had a disproportionately large jaw, Broom himself admits that it would stand between 8 and 10 feet high. Can it not still be called a giant?

The discovery of authentic remains of prehistoric giants has revived a very old riddle, which has always puzzled the learned. There is a passage in Genesis which begins, "There were giants in the earth in those days." And stories of giants—usually man-eaters—are found in the folklore of all the peoples on earth. They are always considered as man's worst enemy and all the purest heroes of legend have a victory over some giant or other to their credit. Skeptics have never failed to counter these legends with the complete absence of tangible remains, until quite suddenly official science came to provide arguments for those who believed in the truth of legend and Biblical infallibility.

Von Koenigswald's and Broom's giants would have left an indelible mark in the memory of ancient man. These huge brutes, not yet entirely graduated from bestiality, must have made life hard for our ancestors. Yet soon these stupid or too peaceable great beasts would have been beaten by man and have gradually fallen back into country where their adversary could not reach them. For giant apes, still agile but no longer able to live in trees, mountains were evidently the most suitable habitat and safest refuge. For the Gigantopithecus which used to live in China, the high Himalayas were the obvious shelter. There, out of reach of their enemies, they could have survived until today, just as their contemporaries have survived in the marshy forests of Borneo and Sumatra. This theory, which is utterly hypothetical, provides the only entirely acceptable explanation of the mystery of the abominable snowman.

I had already related the legends of Himalayan ogres to Von Koenigswald's discoveries for some time when in December 1951 the first photographs of clear footprints of the snowman were published. I took the opportunity to put forward for the first time my view that the snowman was a giant and biped anthropoid no doubt closely related to the Gigantopithecus. At this time scientific opinion almost unanimously rejected the idea that the snowman was an unknown animal. Apart from Sir John Graham Kerr, who gave the *yeti* the benefit of the doubt, and Ivan T. Sanderson, all the zoologists who expressed an opinion tried to identify it with some known animal and nobody took much notice of Ivan and me.

But little by little opinion began to change as the conquerors of Everest brought back the same persistent rumors, Sherpas gave more detailed accounts, a lama claimed to have examined the mummified remains of what he called "big monkeys," and Sir John Hunt himself declared after his interview with the second senior lama at Thyangboche: "There is an interesting problem for an enterprising party to investigate."

By the end of 1953 skepticism was breaking down, and scientists in several countries thought that it was possible that the Himalayan giant might exist and be a new and unknown species. The *Daily Mail* decided to send an expedition to look for the abominable snowman. It set off from Katmandu in January 1954. Besides a mountaineer John A. Jackson, the reporter Ralph Izzard, and Tom Stobart, who made the film *The Conquest of Everest,* the team contained several scientists: Charles Stonor, Dr. Biswamoy Biswas, and Gerald Russell, who took part in capturing the giant panda in 1936.

Charles Stonor had already set off in December 1953 to Namche Bazar, when he met a Sherpa called Pasang Nyima, who said that he had seen a *yeti* three months before some 200 or 300 yards away. It was the first *yeti* Pasang had seen, and was the size and build of a small man. Its head, its body, and its thighs were covered with long hair. Its face and chest looked less hairy and there were no long hairs on the legs below the knees. The color of the fur was "both dark and light" and the chest was reddish. It walked nearly as upright as a man and bent down occasionally to grub in the ground for roots. When it realized that it had been seen it gave a loud, high-pitched cry and ran off into the forest, still on its hind legs, but with a sidling gait. Never once did it go on all fours.

Stonor slyly asked Pasang: "Is the *yeti* a flesh-and-blood animal, or is it a spirit?"

"How could it have been a spirit," the Sherpa replied, "since we saw its footprints after it had run away?"

Stonor collected several other firsthand accounts.

One afternoon in 1947 or thereabouts a yak breeder called Dakhu who lived in Pangboche saw one some 50 yards away. It walked upright, was the height of a small man, and was stocky and covered with hair.

In March 1949 another Pangboche villager called Mingma was alarmed by a *yeti* and took refuge in a stone hut, but was able to see the *yeti* through a large crack in the wall and to describe it to Stonor.

> A squat, thickset creature, of the size and proportions of a small man, covered with reddish and black hair. The hair was not very long, and looked to be slanting upwards above the waist, and downwards below it; about the feet it was rather longer. The head was high and pointed, with a crest of hair on the top; the face was bare, except for some hair on the sides of the cheeks, brown in colour, "not so flat as a monkey but flatter than a man," and with a squashed-in nose. It had no tail. As Mingma watched it, the Yeti stood slightly stooping, its arms hanging down by its sides; he noticed particularly that the hands looked to be larger and stronger than a man's. It moved about in front of the hut with long strides. . . .

When the beast saw that it was being watched through a crack it growled and showed its teeth; Mingma was very struck by their size.

In about 1950 Lakhpa Tensing saw a *yeti* sitting on a rock some 30 yards away. It turned its back to him. It seemed to be the size of a small boy and was covered with rather light reddish hair.

And in October 1952 a villager from Thammu called Anseering and his wife surprised a *yeti* among the rocks when they were going to collect medicinal roots at the upper edges of the forest. The animal was dark brown, smaller than a man and thickset. It made off, climbing over the rocks on four feet.

Continuing his reconnaissance, Stonor finally found some tracks of the *yeti* 14,000 feet up near Namche Bazar. They were the human-looking footprints so often seen by travelers, but smaller than any reported before: the average length was 10 inches, the maximum width 5 inches and across the heel 3 inches. Stonor was much encouraged.

"My own view," he wrote, "is that we are concerned with some quite unknown and extremely interesting beast."

When the main body of the expedition had arrived and were split into three columns so as to search the largest possible area, more *yeti* tracks were soon found. Jackson and Jeeves went up the Khumbu glacier and discovered one, two or three days old, which had undoubtedly been made by a biped. The prints were 10 to 11 inches long by 5 to 6 inches wide.

In the upper Dudh Kosi valley Russell and Izzard came upon a second trail, little fresher than the first. The prints were even smaller, having originally been some 8 to 9 inches long by 4 to 5 inches broad. At the top of Lake Lang Boma they were disconcerted to see that the trail seemed to have been made by a quadruped until it split to go round a boulder, when they realized that it had been made by two bipeds; a second *yeti* had been following in the first's tracks. They followed the trail for some 8 miles and found signs that the animal liked to slide down snowy slopes on its behind, a very playful pastime for a creature that had been called abominable.

Several days later they found a third and then a fourth trail of footprints, so that Izzard wrote to the leader of another column, "You will excuse us if the report of a *single* Yeti's tracks now leaves us rather cold."

Meanwhile members of the expedition also found occasional heaps of excrement along the trails they were following. On his reconnaissance Stonor had twice found a large animal's droppings containing fur, rodents' bones, and earth. The natives that he had asked had all told him that the *yeti* ate marmots and pikas or "mouse hares" and large insects, besides consuming clayey earth "perhaps for bulk or for some mineral value." Some thought it also preyed on young yaks, tahr, and musk deer. Gerald Russell's analysis of *yeti* droppings left no doubt that it was omnivorous:

> a quantity of mouse-hare fur; a quantity of mouse-hare bones (approx. 20); one feather, probably from a partridge chick. Some sections of grass, or other vegetable matter, one thorn, one large insect claw, three mouse-hare whiskers.

Unfortunately although the three columns moved in pincer movements for 15 weeks the *Daily Mail* expedition did not succeed in seeing a single specimen of the elusive animal, at least not for certain, let alone watching it carefully, photographing or capturing it. But they did not draw a complete blank, as Izzard's and Stonor's books prove.

The most interesting information they collected was about the various scalps supposed to belong to the *yeti*. On October 9, 1953, the marks at Pangboche showed a *yeti*'s scalp in the local *gompa* to four Indian mountaineers, Dr. Charles Evans, and Professor Fürer-Haimendorf.

The relic looks like a sort of miter of thick leather shaped like the traditional descriptions of the *yeti*'s head. It is covered with hair 1¼ to 2 inches long on either side of a medial crest of erect hair. The scalp, if it is one, has lost all the hair on the top, not that this means that it comes from a bald *yeti,* but merely that it is an ancient relic which according to the monks is more than three centuries old.

It was photographed and a single hair was given to one of the Indians, who at once sent it to Dr. Hausman, one of the world's greatest experts on the subject.

In 1954 the *Daily Mail* expedition examined the scalp at Pangboche and discovered a second very similar but hairier and apparently less ancient one in the monastery at Khumjung. Several more hairs were removed from the Pangboche scalp and sent for analysis.

All those who examined these two scalps, and they included some experienced naturalists, agreed that they consisted of single pieces of skin and that there was no trace of stitches or glue. But this was not true of a third scalp which the *Daily Mail* team saw in the temple at Namche Bazar, and which proved to be a crude imitation of the other two. It was made of pieces of similar hairy skin sewn together into the required shape.

For three years, starting in 1957, the American oil magnate Tom Slick sent increasingly well-equipped expeditions in pursuit of the snowman. They studied the scalps once again and removed several more hairs to send to the experts. In this way I came by a few.

From December 1959 to February 1960 a Japanese expedition looked for the *yeti* and examined its supposed scalps. Some more hairs were, of course, removed in order to be analyzed under the microscope. This at once showed that all three scalps came from the same kind of mammal. But which?

Before I give the verdict of the experts, I should explain that except in rare cases it is impossible to tell for certain from the study of a hair what group of mammals the animal belongs to. One cannot say, for example, that it must be the hair of a rodent or a primate, a carnivore or an ungulate. The only way of identifying it is by comparing it in turn with hairs of all the likely kinds of mammal. This the experts did, but no animal seemed to fill the bill.

This does not mean that they all thought the scalps were genuine. Far from it. Dr. Hausman thought that they might have been made up out of a skin that had been molded and sewn, and Professor Wood Jones maintained that the hairs came not from a scalp, but from the shoulder of some kind of ungulate.

Now it is undoubtedly true that a flat piece of leather suitably moistened and stretched over a mold can be shaped into a miter. But if the hair tracts are to radiate outward, as they do on the supposed scalps, the piece of leather cannot be cut from any part of the skin. The crest of erect hair seemed to me to prove that the skin of the scalps must have been cut from somewhere along the medial line which runs along a mammal's back from its nose to the end of its tail. Moreover, I had to admit that on a neck or a back the hair always runs parallel to the medial line, whereas in these scalps most of it is at right angles to this line. This arrangement of the hair is found only on the crown of the head. And as a large primate is the only creature with a big cranium of the right shape, I thought this was a conclusive proof that the supposed scalps were genuine.

Then, at the end of November 1960 came the news that Sir Edmund Hillary had been given permission to borrow the Khumjung scalp for six weeks so that it could be examined by specialists in the problem in Chicago, Paris, and London. A fortnight later I finally had the precious specimen in my hands, and my examination of the hair tracts merely confirmed that the supposed scalp did have every appearance of being genuine.

But I began to have my doubts. Hitherto I had only been able to study the excellent black-and-white photographs of the balding Pangboche scalp and a single rather indistinct picture of the one from Khumjung. Now that I could at last examine the latter

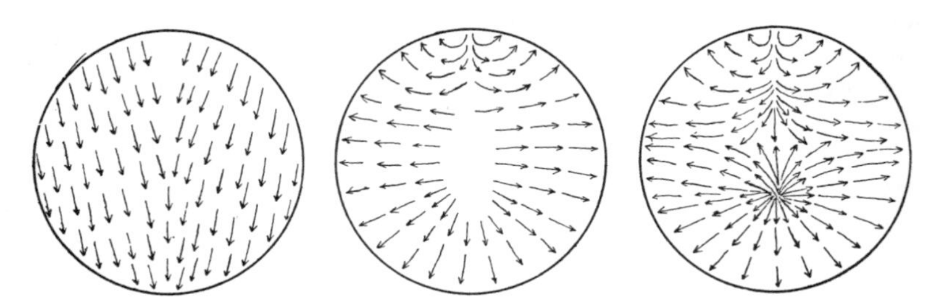

35. The arrangement of hair tracts on (*left*) a quadruped's back, (*center*) the supposed snowman's scalp, and (*right*) a gorilla's scalp. In all three diagrams the animal is facing toward the top of the page.

scalp closely it reminded me strongly of the neck and mane of an animal that I had seen in the Amsterdam Zoo before the war, the southern serow. Now there is in Nepal itself a race of this chamois goat which goes up to an altitude of 15,000 feet in the Himalayas. Thus it is an animal which is found in the same habitat as the *yeti*.

The similarity was striking, but I did not see how the hairs on the back of this goat could be made to radiate outward like those on a scalp, until Ivan T. Sanderson told me he had made several experiments to see whether it were possible to fake *yeti* scalps. He proved to me that if a moistened pelt is much stretched over a hat-shaped mold it produced a sort of "landslide" in the layers of skin. The hair tracts no longer lie at their original angle, but point in the direction in which the skin has been stretched. In this way, when the skin from an animal's neck, with its hair pointing straight backward, is stretched over a mold shaped like the nose of a shell, the hair points outward like that on the head of a primate, whether man, ape, or ape man.

I had now merely to check whether my impression was correct, by comparing my samples of hairs from the various scalps with the skin of a Nepalese serow. But where could I find a specimen which was so rare that there was not even one in the Paris Natural History Museum?

In the end I discovered that there was a mounted specimen in Brussels. A comparison of its hair with those from the supposed scalps clinched the matter; they clearly all came from the same species of animal. No further doubt was possible: the snowman's scalps did not come from the snowman and were not scalps. They were more like wigs made by stretching the skin from the neck of the Nepalese serow over a mold. And because this animal is so rare in our collections, the greatest experts on hair did not compare the scalp hairs with it and were never able to identify them.

The fact that the scalps are not genuine does not affect the main problem. The *yeti*'s existence is not based solely on these wigs. Because "leopard skin" is now made of nylon fur, it does not mean that leopards do not exist; it is, rather, evidence that they do! But why on earth should the lamas want to make fake *yeti* scalps?

It has been known for some time that the lamas at Namche Bazar had faked one rather crudely by sewing together several bits of skin, perhaps because they were jealous of the prestige which the lamas of Pangboche and Khumjung gained from their genuine scalps. But in fact both these scalps were not considered as equally genuine. My friend Peter Byrne, who went on Tom Slick's expeditions, wrote to me on January 24, 1961, that it was notorious all over northern Nepal that the Khumjung scalp was a fake and had been made some twelve to fifteen years before by a Tibetan taxidermist so the lamas could keep up with their fellow monks at Pangboche.

On the other hand, everybody thought that the old Pangboche scalp was genuine. And it was probably not a deliberate fake. In their ritual dances the Buddhist priests represent gods, demons, and animals both legendary and real. For this purpose they use not only masks and costumes but also animals' skins, shells, horns, and antlers. The miters made of serow skin shaped to resemble the pointed head described by the witnesses of the *yeti* are worn by the dancers representing this animal. The Pangboche scalp is pierced by a series of holes all along one side and also near the top. Through the lowest holes were threaded the laces to fasten it to the dancer. The holes near the top were probably used for attaching silk prayer flags. In the course of the centuries what was originally merely a theatrical property came to be thought of as a genuine piece of the animal's skin.

Ever since the *Daily Mail* expedition the *yeti* has been in the news. In May 1955 the French expedition on Makalu came upon several tracks. The Abbé P. Bordet, the team's geologist, who photographed the imprints, wrote:

> I have followed a *yeti* track for more than a kilometer, and seen nearly 3,000 footprints. They are all of the same kind. They are deep marks made by a foot somewhat resembling a human foot. The sole of the foot is roughly elliptical and rounded underneath. In front of it are the more or less circular marks of four toes (not five), the first on the inside is larger than the rest and perhaps not quite so far forward, the other three lie on the front edge of the sole of the foot and very close to it. These toes are much larger than human toes. There are no marks of claws. . . . In the best imprints there are still little ridges of snow dividing the toe-marks and showing that the toes are slightly separated when the creature walks. The length of the footprints is about 20 centimeters. . . .

"Whether it is a kind of bear or ape," Bordet concludes, "it seems too soon to decide." But since he tells us that "the first [toe] on the inside is larger than the rest," the *yeti* cannot possibly be a bear, even if it is an unknown biped species. He also points out the odd fact:

The Indian map of the Himalayas marks the area round Everest as the Mahalangur Himal (the mountains of the great monkeys). As no monkey is known to live there this name may refer to the *yeti*, a characteristic inhabitant in the natives' eyes.

A month later, on June 12, 1955, two members of the Royal Air Force Mountaineering Association expedition to the Himalayas also found fresh *yeti* tracks 12,375 feet up in the Kulti valley.

> There were many prints, each measuring about 12 inches by 6 inches, and indicating that the creature who made them was two-legged, with five toes a quarter of an inch wide on each foot.
> The prints were sunk 11 inches into the snow, compared with the one-inch impression made by the R.A.F. mountaineers.

This shows that the creature which made such deep tracks must have been several times as heavy as a man. (A gorilla, by the way, rarely exceeds 560 pounds.)

Professor René von Nebesky-Wojkowitz, who spent three years in Tibet and Sikkim, tells several stories about the abominable snowman in his admirably documented book. He writes:

> It is a remarkable fact that the statements of Tibetans, Sherpas and Lepchas concerning the Snowman's appearance largely coincide. According to their description a warrant for the arrest of this most "wanted" of all the inhabitants of the Himalayas would read as follows: 7 feet to 7 feet 6 inches tall when erect on his hind legs. Powerful body covered with dark brown hair. Long arms. Oval head running to a point at the top, with apelike face. Face and head are only sparsely covered with hair. He fears the light of a fire, and in spite of his great strength is regarded by the less superstitious inhabitants of the Himalayas as a harmless creature that would attack a man only if wounded.

He adds some more interesting information about the creature's habitat:

> From what native hunters say the term "snowman" is a misnomer, since firstly it is not human and secondly it does not live in the zone of snow. Its habitat is rather the impenetrable thickets of the highest tracts of Himalayan forest. During the day it sleeps in its lair, which it does not leave until nightfall. Then its approach may be recognized by the cracking of branches and its peculiar whistling call. In the forest the *migo* moves on all fours or by swinging from

tree to tree. But in the open country it generally walks upright with an unsteady, rolling gait. . . . The natives . . . say the Snowman likes a saline moss which it finds on the rocks of the moraine fields. While searching for this moss it leaves its characteristic tracks on the snowfields. When it has satisfied its hunger for salt it returns to the forest.

This suggests that—as I suspected—the snowman owes the accidental fact that it is a biped to snow beneath its feet, and also that the places where the creature and its footprints are usually seen are not its normal habitat.

The first American expedition led by the Texas oilman Tom Slick also searched for the snowman in eastern Nepal in the spring of 1957. Not only did he examine three sets of footprints, excrements, and even hairs that had come from *yetis,* but he also questioned some 15 Nepalese who claimed to have seen the animal. His most important contribution to the problem is, I think, his belief that there are two kinds of *yetis*: one with blackish hair and about 8 feet high, and the other reddish and smaller. This conclusion is based on differences in the accounts he heard from the natives, on finding two kinds of hair along the tracks and examining two kinds of footprints. The largest were as much as 13 inches long; the smaller looked like those photographed by Shipton but were about the size of a man's.

This would explain the discrepancies that the reader cannot fail to have noticed in this chapter. While some witnesses say the snowman is a giant, all the Sherpas who have seen it say that it is smaller than a man or about the same size. It is significant that all the reports which say the *yeti* is a giant come either from Tibet or from the very north of Sikkim or Nepal; that is, from very high places on the edge of these countries. I should also mention that a teacher at Namche Bazar told Ralph Izzard that there were two kinds of *yetis.*

This opinion was confirmed in 1957 by a Tibetan lama called Punyabayra, who said that the Tibetan mountain people knew three kinds of snowmen: the *nyalmo,* the *rimi,* and the *rackshi bompo.* The *nyalmo* are real giants, between 13 and 16 feet high, with enormous conical heads. They wander in parties among the eternal snows above 13,000 feet and are carnivorous and even man-eating. The *rimi* are smaller, but still between 7 and 9 feet high. They live lower down, between 10,000 and 13,000 feet, feeding on plants as well, and are thus omnivorous. Their favorite place is the Barun Khola valley in eastern Nepal. The *rackshi bompo* are about the size of a man.

Some may say that these are the absurd superstitions one might expect a Tibetan lama to spread. But Punyabayra is an educated man, and his account agrees in part at least with what we already know. The *rackshi bompo* must be the Sherpas' reddish *yeh-teh* or *mi-teh* which leaves the footprints 8 or 9 inches long that the *Daily Mail* expedition and the Abbé Bordet found in such quantity. The *rimi* must be the large black snowman described by Nebesky-Wojkowitz's Tibetan and Sikkimese informants, and the one whose mummified bodies Chemed Rigdzin Dorje saw in Tibet, and which leaves footprints 12 to 13 inches long. It was in fact in the Barun area or in eastern Nepal that Shipton and Slick found footprints of this size. But do the *nyalmo* really exist, or are they just a myth? The evidence is far too slender for us to draw any satisfactory conclusions. Possibly they are an invented addition based on the belief that *yetis* increase in size the higher you go.

Meanwhile we can say that the existence of two kinds of *yetis* of different size and color is well established. They may be different geographic races of the same species; or merely the result of sexual dimorphism; or the small reddish *yetis* may be the young of the large black ones.

We can now even give a detailed description of the giant biped anthropoid of the Himalayas, a survivor of the giant primates which once ruled a large part of the earth. Only four are so far known to paleontology: the Chinese Gigantopithecus, of which we have only a few molars and a jawbone, the Java Meganthropus, represented by a fragment of jawbone with its teeth, the Tanganyika Meganthropus, consisting only of jaw and facial bones, and the South African Paranthropus, of which several skulls and a biped's pelvis have been excavated. If I seek to relate the snowman to the Gigantopithecus it is obviously for geographical reasons and because most reports of its size agree with Dr. Broom's estimate of that of the Chinese giant.

It is high time that the snowman was also recognized by a scientific name and since the paleontology of giant primates is so slender, I would give it a new name, *Dinanthropoides nivalis*, or "terrible anthropoid of the snows." If one day its teeth are examined and found to be identical with those of the Gigantopithecus, its name will have to be changed to *Gigantopithecus nivalis,* the present species being no doubt quite distinct from the Pleistocene primate from Kwangsi.

We have reason to believe that it is a large biped anthropoid ape, from 5 to 8 feet high according to its age, sex, or geographic race, which lives in the rocky area at the limit of the plant line on the slopes of the whole Himalayan range. It has plantigrade feet, and the very conspicuous big toe, unlike that of most monkeys, is not opposable to the other toes. It walks with its body leaning slightly forward; its arms are fairly long and reach down to its knees. It has a flat face, a high forehead, and the top of its skull is shaped like the nose of a shell; its prognathism is slight, but its thick jaws have developed considerably in height, hence the disproportionate size of its molars. To this outsize masticatory apparatus are connected very powerful jaw muscles. On the cranium there is a sagittal crest which is revealed by a thickening of the scalp in the adult male, at least, and the presence of upstanding hair. It is covered with thick fur, which in the smaller specimens varies from fawn to dark chestnut in different places with foxy-red glints, but the face, chest, and the lower legs are much less hairy. In the larger specimens the fur is an even darker brown or almost black. It appears to be omnivorous: roots, bamboo shoots, fruit, insects, lizards, birds, small rodents, and occasionally larger prey like yaks are all grist to its mill in such barren country. Its cerebral capacity should be about equal to or even greater than man's.

Of course it is not until an actual specimen of the snowman is examined that my deductions can be checked and its provisional description completed. The chances of success seem to be seriously compromised for the future. Slick had no sooner announced his intention of returning to the Himalayas in the autumn of 1958 when the Government of Nepal announced that it required a fee of 5,000 rupees from anyone wishing to look for the *yeti*. It strictly forbade anyone to kill one except in self-defense. The Nepalese were not allowed to give information to strangers without permission from the Government, which claimed a right in the creature, dead or alive, and even in any photographs that might be taken.

This decree has not discouraged the enterprising. Slick, as he had planned, sent a second expedition to Nepal in 1958 and a third in 1959, when there were also two Japanese expeditions on the track of the snowman.

Meanwhile the *yeti* had become news in Russia.

Some Marxist diehards had maintained that the snowman was pure invention, a pretext to enable Western expeditions to spy on the frontiers of Nepal and Tibet. But in January 1958 a Soviet scientist, Dr. Alexander G. Pronin, announced that he had seen one the previous August in the Baliand-Kiik valley in the Pamirs, thus extending the animal's area of distribution enormously.

> At first [he told *Komsomolskaya Pravda*] I thought it was a bear, but, having collected myself, it became clear that this was no bear but a manlike creature. It walked out of its cave for a distance of some 200 yards and then disappeared beyond the edge of the cliff.

He had observed the creature for some five to eight minutes, and three days later he saw it again at the same spot. He described it as "a manlike creature walking on two feet in a slightly stooping manner and wearing no clothes. Its thickset body was covered with reddish-gray hair, and it had long arms."

Soon after Dr. Pronin's announcement a Chinese photographer confirmed that there were snowmen in the Pamirs. In 1954 when he was filming three colleagues on Mount Muztagh-Ata, 20,000 feet up, he saw two of these creatures one morning.

Dr. Pronin's report was much criticized in Russia. But many first-rate Soviet scientists, the late zoologist P. Sushkin, Professor N. Sirotinin, Professor S. Obruchev, and the greatest of their zoologists, E. N. Pavlovsky, have, however, all come out on the side of the snowman's existence.

Since 1957 a commission set up by the Soviet Academy of Science at the instigation of my good friend Professor Boris F. Porshnev has been studying the snowman and has undertaken the enormous task of collecting all the reports of unknown hairy biped primates throughout Asia and publishing them in a series of pamphlets. These have revealed that the area in which similar creatures have been seen is extraordinarily large: in Asia it spreads from the Sayan mountains and the Gobi desert in the north to the Himalayas and Assam to the south, from the Caucasus in the west to the Mhinghan, Tsinling-Shan, and Nan-Shan mountains in the east.

Moreover, since 1959 there have been reports of creatures like large *yeti* on the other side of the Atlantic in the pine-forests on the Klamath Mountains in the north of California. I have myself found evidence of several types of similar creatures in Africa. If the reader wants to study the problem in all its complexity, I can strongly recommend Ivan T. Sanderson's exciting, daring, and very well-documented *Abominable Snowmen* (1961), in which he makes use of all my own files on the subject containing a great deal more information than I had when this book was first published.

Actually it is not absolutely necessary to examine a living *yeti*—nor to kill a specimen, which would be needlessly cruel—to clear up the problem once and for all. The lama Chemed Rigdzin Dorje may be right when he says that some monasteries possess mummified snowmen. A study of one of these specimens would settle the question. At Pangboche there is a mummified hand that is supposed to be a snowman's. It has been photographed from all angles by Tom Slick's team, and the resulting pictures have been studied by a number of zoologists and anthropologists. Moreover, a piece of dried skin, a joint of the index finger and the whole thumb have also been submitted to their examination. The experts are not all of the same opinion, but they do all agree in thinking that the hand is not quite like the hand of *Homo sapiens*. "Almost human," "perhaps half-way between that of an anthropoid and a man," "an abnormal human hand" and "similar in some respects to that of Neanderthal Man" are some of the opinions they have expressed.

The expedition which may one day succeed will be able to benefit from the lessons of the *Daily Mail* team, who never managed to overtake a *yeti* although their trails abounded. Each time they lost track of the strange ape man, which could move faster over a terrain to which it was better suited. And, as Izzard remarked, "a party is as conspicuous as a line of black beetles on a white tablecloth." They would probably have done better to follow Gerald Russell's plan. He was more expert in catching animals than anyone else in the party, and he thought that rather than keeping on the move, they should have chosen a good strategic position and remained hidden there, if necessary for a week at a time. As it was, Izzard was a sadder and a wiser man for this experience, and concluded:

> The Yeti is more likely to be met in a chance encounter round, say, a rock, than by an organised search. A reconnaissance party of two or three Sahibs needs about 30 Sherpa porters to support it over a period of about three weeks. We found it impossible to introduce such a large body of men into an "empty quarter" of the Himalayas without disturbing all wild life within it.

We shall find the same situation recurring in most of the problems of unknown animals, and it is this that makes them so hard to solve. It is difficult not to be exasperated when all the pieces of evidence run away as soon as the experts arrive on the scene.

PART THREE

THE LIVING FOSSILS OF OCEANIA

Forth from the dead dust, rattling bones to bones
Join; shaking convuls'd, the shiv'ring clay breathes,
And all flesh naked stands:

WILLIAM BLAKE, *The Song of Los.*

BORNEO
CELEBES
MOLUCCAS
NEW GUINEA
Banda Sea
JAVA
Bali
Lombok
Sumbawa
Komodo
Flores
Rita
Pantar
Timor
Timor Sea
NEW GUINEA
Darwin
Arnhem Land
Gulf of Carpentaria
Cape York Peninsula
Coen
Great Barrier Reef
Cairns
Trinity Bay
Normanton
Tully
Rockingham Bay
Cardwell
Atherton Tableland
Dalrymple
NORTHERN TERRITORY
QUEENSLAND
Yamba
Australian Cordillera
Great Sandy Desert
Alice Springs
WESTERN AUSTRALIA
Salt Lake
Shark B.
SOUTH AUSTRALIA
L. Eyre
Mundubbera
Fraser I.
Tiaro
Dalby
Brisbane
St. George
Callabonna
Great Victoria Desert
Nullarbor Plain
Darling R.
NEW SOUTH WALES
Perth
Swan R.
Crystal Brook
Gloucester
Wellington
Murray R.
Orange
Newcastle
Leeton
Narrandera
Bathurst
Adelaide
Murrumbidgee
Harden
Sydney
L. Alexandrina
Riverina
George
Jamberoo
Kangaroo I.
Conargo
Goulburn
VICTORIA
Chiltern
Wagga Wagga
Shoalhaven R.
Lockwood
Euroa
Moruya
Tantanoola
Mansfield
Eumeralia R.
L. Burrumbeet
Ballarat
Gippsland
Briagolong
Melbourne
Port Fairy
Corangamite
Colac
Port Phillip
AUSTRALIA
Bass Strait
TASMANIA
Great Lake
L. Tiberias
Hobart
Miles
0
200
400
600
Tropical Forests

CHAPTER 7

THE INCREDIBLE AUSTRALIAN BUNYIPS

I shall laugh myself to death at this puppy-headed monster.

SHAKESPEARE, *The Tempest.*

AT THE BEGINNING of the seventeenth century, the brave seamen sent out by the Dutch East India Company several times set foot on what seemed to be a very vast new land, which they loyally named New Holland. There they said there lived an animal as large as a man, with a head like a deer and a long tail. It stood on its hind legs like a bird and could hop like a frog. In 1640 Captain Pelsart gave an accurate description of it, only to be greeted with derision at the tall stories told by seamen. In 1770 Captain James Cook landed in Trinity Bay to repair the *Endeavour,* which had been damaged on the Great Barrier Reef. On July 9, Sir Joseph Banks, the famous naturalist and patron of the sciences, went off with one companion to look for game to replenish their rations. In Cook's words:

> Going in pursuit of game, we saw four animals, two of which were chased by Mr. Banks's greyhound, but they greatly outstripped him in speed, by leaping over the long thick grass, which incommoded the dog in running. It was observed of the animals that they bounded forward on two legs instead of running on four.

Later Cook learned that this sort of giant jerboa was called *kangaroo.** His evidence had to be believed. But it was twenty

* It is often said that this word does not refer to the animal in question but merely represents the puzzled natives' reply to the English explorers' questions. Some say that it means "I don't understand"; others that it is a garbled repetition of the beginning of the question, "Can you tell me . . . ?" These stories are a complete fabrication.

years before Dr. George Shaw gave the kangaroo its Latin name. Cook had overlooked its strangest feature: the pouch on the female's belly in which she carries her young. Marsupial pouches were not new to zoologists. It had long been known that the opossum had a similar pouch. But they were so convinced that it was peculiar to the opossum that when Seba described an "eastern philander" or cuscus, a pouched animal that had been sent him from Amboina, Buffon thought that there had been a mistake.

Soon it was found that *all* the indigenous Australian mammals had pouches. What was the rarest exception elsewhere was the rule here. More surprises were to follow. In 1797 a settler in New South Wales found an animal which he called an "aquatic mole." Actually it was more like an otter, but it had a duck's webbed feet and, what was even more extraordinary, a duck's beak. When the first skin arrived in London and was examined by the Zoological Society it was almost unanimously thought to be a fake, like the confected mummies that travelers brought back from China. Nevertheless Dr. George Shaw examined the skin minutely and found no trace of glue or stitches. He announced that the skin was genuine and in 1799 published the first scientific description of the platypus.

It was one thing to give the strange animal a name, it was quite another to classify it in the system of zoology. As it was covered with hair, Shaw had no doubt that it was a mammal. But in which order did it belong? Blumenbach put it in the Edentates, that rag bag of unclassifiable mammals. Then in 1802 two more specimens, this time preserved in alcohol, arrived in England. One of them was a female, and the zoologists were astonished to find that this presumed mammal (or "animal with teats") hadn't

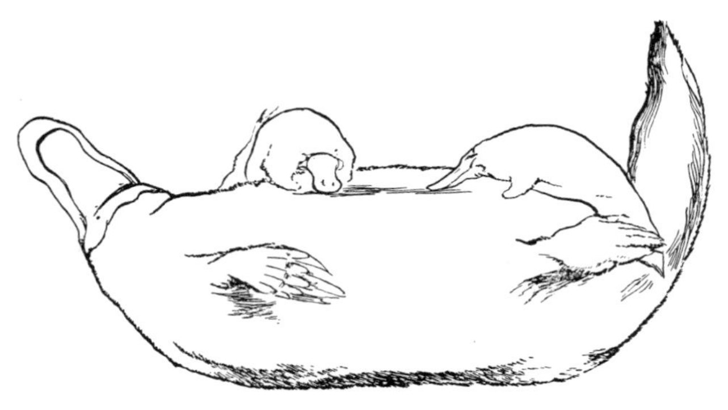

36. Platypus suckling its young (after Hartwig, 1910).

any teats at all! Moreover neither of them had a separate anus and genital duct like the rest of the mammals, but a single cloaca like the reptiles and birds. Yet they could not belong to either of these classes since they were covered with fur, not scales or feathers. The anatomist Everard Home therefore proposed to create a separate order to include this duck-billed animal with another creature with a cloaca which had been discovered in Australia in the meantime: the echidna, a sort of porcupine whose long snout also looked like a beak. This was done a year later when Étienne Geoffroy-Saint-Hilaire created the new order of monotremes or "animals with a single hole." But he prudently refrained from stating which class this order belonged to.

Then news arrived from Australia that the platypus laid eggs, or so Sir John Jamison wrote on March 17, 1817. This brought it even closer to the birds and reptiles; it certainly could not be a true mammal. But in 1824 the German scientist Meckel, examining the animal more carefully, found that it had mammary glands. They had not been noticed before because they were very small except in the breeding season.

Impossible! argued Étienne Geoffroy-Saint-Hilaire and his son Isidore, supported by Blumenbach. If the monotremes lay eggs—and they do lay eggs—they cannot have mammary glands. The glandular formations must be musk glands.

No, replied Meckel, they are true mammary glands. The monotremes are true mammals and they are therefore certainly viviparous. The legend that the platypus lays eggs is absurd.

And such eminent scientists as Cuvier, Oken, and Blainville agreed with him. There were others, like Richard Owen and Home, who subtly sat on the fence and averred that they must be ovoviviparous—that is to say, they formed eggs which hatched inside the body and so the young were born alive.

In 1829 Robert E. Grant claimed he had found four platypus eggs and sent a very careful drawing of them to Europe. The triumph of the Saint-Hilaire clan was short-lived when experts identified the eggs as those of an Australian long-necked tortoise.

In 1832 Meckel, Cuvier and Co.'s victory seemed complete when an Australian naturalist, Lieutenant Maule, established without a shadow of a doubt that the glands on the platypus's belly actually produced milk. There was a small fly in the ointment, however: Maule had found eggshells in the platypus's burrow. But these eggs could have—indeed they *must* have—come from some other animal.

Some thirty years later, on September 21, 1864, Dr. Nicholson sent a letter to Professor Owen in which he told how a platypus that a workman had caught and put in a gin crate laid two eggs during its first night in captivity. The Professor was not convinced. He had stated once and for all that the monotremes were ovoviviparous. The platypus must have had a miscarriage from fright.

It was not for another twenty years that the matter was settled. The reproduction of the monotremes had been disputed for almost a century when on the same day in 1884 two naturalists—Dr. W. H. Caldwell, who was working on the platypus, and the German Professor Wilhelm Haacke, who was studying the echidna—simultaneously announced that they had proved that the monotremes were oviparous.

Everybody had been wrong.

I have related the platypus's history at such length in order to show that an animal which all the scientists think impossible may actually exist. The *bunyip* is not half so impossible as the platypus.

What is a *bunyip?* Nowadays the Australians use it to mean a "bogey," something at once frightening, shadowy, imaginary, and rather ludicrous. But to some men it is a flesh-and-blood animal with a clearly defined appearance and habits.

The earliest report comes from the French explorers in the crew of the *Géographe.* In June 1801 they made their way into the interior, having named a bay on the southern tip of the west coast after their vessel. Suddenly they heard a terrible roar, louder than a bull's bellow, which seemed to come from the reeds in the Swan River. They were terrified and took to their heels, convinced that it was some great water beast. Then the explorer Hamilton Hume said that in Lake Bathurst, at the other end of the continent, he had seen an animal like a manatee or hippopotamus. Five scientists of the Philosophical Society of Australasia at once promised to pay all his expenses if he managed to procure a skull, a skin, or a skeleton of his unknown animal. Hume never found one, but he had not invented the story, for many similar tales at once began to come in from various places in Australia, especially in the southeast.

By the middle of the nineteenth century the legend of the *bunyip* was well established. In 1846 a fragment of a skull was found on the banks of the Murrumbidgee, a tributary of the Murray River. It was sent to the naturalist W. S. Macleay as the skull of a "true *bunyip*." He identified it as the skull of a deformed colt. But when Professor Richard Owen saw a drawing of it in London he immediately decided that it was a calf's. One of the experts must be wrong. So the problem was not so puerile as they seemed to imply. Quite possibly they were both wrong. We shall never know; the specimen was deposited in the Australian Museum, but, as one might expect of a *bunyip,* it has mysteriously disappeared.

In 1847 Governor Latrobe managed to get hold of copies of native drawings of two kinds of *bunyips* from Victoria. He sent them to Tasmania, but, alas, they too seem to have vanished.

In 1848 one turned up in the Eumeralla River in Victoria. It was said to be a large brown animal with a head like a kangaroo. It had a long neck with a hairy mane, and its mouth was enormous.

In 1872 a strange "animal like a big retriever dog, with a round head and hardly any ears" came so close to a punt in Lake Corangamite that the terrified man on board capsized his craft.

The *wee-waa* which was seen in 1872 and 1873 in a lagoon near Narrandera was also compared to a retriever by those who saw it; but it was only half the size, and its body was covered with long hair, shining and as black as jet. In 1873, also, a creature with a seal's head was supposed to have been seen rising out of the water near Dalby in Queensland. It had a double but asymmetrical caudal fin. In 1876 a hairy animal of unusual appearance turned up in Crystal Brook in South Australia. Reports continued to come in and most of them sounded like distorted descriptions of a seal or a dugong.

But not all were as sober as those I have just quoted. The Reverend George Taplin described the *moolgewanke,* or *bunyip* of Lake Alexandrina, as a weird creature, half man, half fish, with hair like a sort of wig of reeds. It had a voice like the rumble of a distant cannon and gave you rheumatism.

The *tunatpan* of the Port Phillip area was even more picturesque and recalled the Eumeralla *bunyip.*

> It was as big as a bullock, with an emu's head and neck, a horse's mane and tail, and seal's flippers, which laid turtle's eggs in a platypus' nest, and ate blackfellows when it was tired of a crayfish diet.

In 1886, horsemen fording the Molonglo River saw a creature, "whitish in colour and about the size of a large dog. Its face was like the face of a child." All the same they had no qualms about throwing stones at it and driving it away. The *bunyip* of the Tuckerbil swamp near Leeton which was reported until the summer of 1929–30 was frankly marvelous. It was "a two-headed Bunyip which could swim both ways without changing gear." But other reports must have had more of a ring of truth, for in 1890 a team from the Melbourne Zoo made a systematic attempt to capture a *bunyip*-like monster in the Euroa district.

Similar water monsters have several times been reported in Tasmanian lakes. In the autumn of 1852 one was seen in Lake Tiberias. It was between 4 feet and 4 feet 6 inches long, had a bulldog's head, and was covered with shaggy black hair. No doubt they were similar animals which were often seen in the Great Lake. The length was always between 3 and 4 feet and the head was round like a bulldog's. Often they splashed water 6 to 10 feet into the air. Between 1860 and 1870 there was no lack of reports about them in the same terms. In 1863 someone was able to observe one for some time and described it as a sort of sheepdog with two little fins, like small wings. It moved at the considerable speed of about 30 miles per hour. *Bunyips* reappeared several times in 1932.

"Apart from the most extravagant fabrications," Gilbert Whitley of the Australian Museum concludes, "we must be struck with the comparative uniformity of bunyip description over a long period." Generally it is a large furry animal, the size of a dog and with a dog's head, but with very small or even invisible ears. It swims with fins and lives in rivers, marshes, and inland lakes.

This description reminds one of a seal or a sea lion; and there are several genera of these animals in Australia and Tasmania. But could these sea animals reach inland waters? There is no reason why they should not, for a species of seal exists which lives only in the lakes and inland seas of Asia. Moreover in Australia seals and sea lions are said sometimes to reach the drainage basin of the Murray River and its tributary, the Darling. Dr. Charles Fenner mentions a seal killed at Conargo, 900 miles from the mouth of the Murrumbidgee. A sea leopard was killed in the Shoalhaven River in 1870. When it was cut open an adult platypus was found in its stomach. As Whitley remarked, "Surely a bunyip within a bunyip!"

Thus it seems well established that seals can travel long distances in fresh water, and it is possible that they may go a short way over dry land. Most of the reports about "water devils" come from the southeast of Australia; that is to say, from the drainage basins of the only two large rivers on that continent.

But many reports of *bunyips* come from places that seals and sea elephants could not reach. Australian scientists have therefore put forward much bolder theories to explain its presence.

> The Bunyip [writes Whitley] has been thought to have been an extinct marsupial otter-like animal, rumours of whose existence have been handed down in aboriginal legends, the latter corrupted and confused with crocodiles in the north of Australia and seals in the south.

Why "extinct"? The reports that I have mentioned speak of a living animal, not a ghost. There are marsupial equivalents of most types of placentary mammals: running, jumping, climbing, tree-dwelling, gliding, and burrowing types. "Nevertheless," as Jean Piveteau has remarked, "certain types of adaptation are missing; there are none truly aerial or truly aquatic." Might not the *bunyip* be an example of the missing aquatic type or an unknown species of platypus? In any case the insistence of some zoologists that legends of unknown animals are based on extinct species seems to be an *idée fixe*. Thus Dr. Anderson, Director of the Australian Museum, writes, "There may be aboriginal legends of the former existence of the *Diprotodon,* which is commonly believed to have haunted rivers, swamps and inland lakes." But that is another story.

When Mary Chase wrote *Harvey,* a play about a drunk whose best friend is a rabbit 10 feet high that no one else can see, she must have chosen this eponymous and invisible animal for its very improbability. Yet there are plenty of gold prospectors who have returned from their gold-panning expeditions in the vast interior with ludicrous tales about rabbits 10 feet long. Naturally there were hints that these monsters, like Harvey himself and the pink elephants so beloved of humorists, were a by-product of alcohol; for gold mining is not an abstemious trade.

But these stories aroused the interest of some scientists, among them Ambrose Pratt, the great Australian naturalist. He seriously asked whether these Brobdingnagian rabbits might not be surviving specimens of the Diprotodon, a gigantic wombat, as large as a rhinoceros, which once wandered the Nullarbor Plain, before the increasing drought turned most of the continent into a desert. Their skulls are as much as 3 feet long. There is no lack of information about their anatomy, for more than a hundred more or less complete skeletons were discovered in the Salt Lake at Callabonna.

This creature's fat body reminds one more of a huge bear than of a rabbit; but, like the kangaroos, it has a face very like a rabbit's or a hare's, with two conspicuously large incisors in the upper jaw under a typical "harelip," hence the name of Diprotodon ("with two teeth in front"), a characteristic of the vegetarian mar-

37. Reconstruction of the Diprotodon.

supials. But, whether the Diprotodon looked like a rabbit or not, could it have survived until today?

The comparatively recent climatic upheavals which have transformed the appearance of Australia are usually blamed for the Diprotodon's extinction. It is supposed to have enjoyed a semi-aquatic habitat among the lush vegetation which still covered the continent at the end of the last Ice Age, between 12,000 and 30,000 years ago, until the ever-increasing drought ate away the country like a leprosy.

No doubt the Diprotodon tried to survive in various oases, and, as they dried up, vast hordes must have set off to wander in search of new water holes until they perished of thirst. In June 1953 Professor Ruben A. Stirton of the University of California returned to Adelaide from the dry northwest of South Australia where he had found a Diprotodon graveyard containing between 500 and 1,000 skeletons in a superb state of preservation. A vast herd of these poor beasts, dying of thirst, had apparently ventured onto the hard crust where a lake had recently dried up to no more than a few tempting pools. The crust gave way under their weight and they gradually sank into the soft clay, which preserved their skeletons from the weather. Some of them had their paws folded beneath them, as if they had met their death calmly welcoming the moisture of their muddy grave.

How long can one say for certain that the Diprotodon survived? Among some aboriginal tribes legends are still current which seem to refer fairly exactly to them. Professor John Walter Gregory is firmly convinced that the central Australian aborigines' mythical *kadikamara* is based on knowledge of living Diprotodon. And in 1872 Dr. George Bennett, a paleontologist of the Australian Museum, reported that "Charlie Pierce," an aborigine, had told him that certain fossils were those of an animal, long extinct, known to the natives by the name *gyedarra*. It was as large as a heavy draft horse, walked on all fours, ate grass, and spent most of its time in the water.

Now Van Gennep considers that an oral tradition cannot originate more than two or three centuries ago, without being radically distorted and transformed. Thus even the boldest estimates, according to which the Diprotodon still haunted the central plains hardly 2,000 or 3,000 years ago, would seem to be too cautious, and one should take the more heed of prospectors' tales of "giant rabbits." For the drought which certainly decimated the Diprotodon need not have exterminated them entirely, any more than it has the other vegetarians—and the carnivores that feed upon them—that are found in the continent. Moreover, there are still lakes, marshes, and streams where the Diprotodon might find shelter as other animals do.

But the "giant rabbits" have, according to the prospectors' tales, a disconcerting habit. Witnesses have often stressed the extraordinary speed at which the "rabbits" made off. Indeed, these giants may have been likened to rabbits because of their sudden flight as much as their general appearance. Yet it is hard to imagine a Diprotodon running like lightning. Are not these "rabbits" more likely to be a gigantic form of kangaroo, which can vanish in one jump? The larger known species can jump 30 feet in one bound. And some reports have referred explicitly to "kangaroos 12 feet high." The discovery of fossil remains of similar giants (*Paleorchestes*) has given added weight to these rumors, especially as it is well established that these animals lived recently.

Whether it is a Diprotodon or not there is little doubt that an animal of unusual size lives in the deserts of Western Australia, but what it is still remains to be settled. Even the boldest naturalists hesitate to venture into this hostile country, remembering perhaps the tragic fate of Dr. Ludwig Leichhardt. He arrived in

Australia in 1841 and immediately went right across the whole continent to Fraser Island, covering some 2,500 miles. He tirelessly carried out several other explorations, and his researches gradually led him to the belief that the Diprotodon was not extinct. In 1847 he organized an expedition with the secret hope of bringing one back alive. He set off enthusiastically into the dry heart of the continent and it is known that he reached the River Cogun on April 3, 1848. Nothing more was ever heard of him or his companions. The trail of unknown animals sometimes leads to Hell.

CHAPTER 8

THE QUEENSLAND MARSUPIAL TIGER

"You'll see me there," said the Cat, and vanished.
Alice was not much surprised at this, she was getting so well used to queer things happening.
LEWIS CARROLL, *Alice in Wonderland.*

TO THE ZOOLOGIST Australia is indeed an upside-down country. While the animals found there were strange, if not extraordinary—the marsupials, and especially the monotremes, exceeded the scientists' wildest hopes—explorers and settlers constantly but vainly looked for familiar animals, as if to reassure themselves that this bewildering place was somehow normal.

The marsupial tiger is the most important of these apparently familiar animals to be found in Australia. Its existence is no longer in doubt—though it is still not officially recognized—but the first reports were based on misinterpreted rumors.

In November 1642 Tasman sighted land which he took to be the southern tip of New Holland. He named it Van Diemen's Land in honor of the Dutch governor on whose orders he was seeking a southern sea route between America and the Orient. It was none of his business to explore new lands, and he did not set foot on the island that was later to bear his name; but he sent his chief pilot, Visscher, ashore with a handful of men. They came back with some alarming tales. Besides notches in trees which they thought had been made by giants they had seen footprints with claws like a tiger's.

There is no doubt that these had been made by a Tasmanian or pouched wolf (*Thylacinus cynocephalus*), a sort of marsupial wolf, which, by an odd chance, is striped like a tiger on the middle of the back, rump, thighs, and tail. It has five toes on its forefeet, the footprints of which cannot therefore be confused with a dog's or cat's. It has only four toes on its hind feet like a dog or cat, but

38. Thylacine or Tasmanian pouched wolf.

whereas it walks on its fingers, it puts the whole sole of the foot on the ground, and so leaves a relatively larger footprint. Hence the confusion with tigers. Otherwise this creature is much more like a dog—its Latin name means "the pouched dog with the wolf's head." And although fossilized remains have been found on the Australian mainland, it has long been confined to Tasmania.

What then is the origin of the tales of tigers—and even lions—that have been current all over Australia since it was first colonized. R. W. Mackay writes:

> There are records from Pipers Creek, Mansfield, Lockwood, Chiltern, Briagolong, and other places in Victoria; from Harden, Tantawonglo, Goulburn, Gloucester, Wellington, Jamberoo, Orange and other places in New South Wales; from the Three-Mile Scrub, Brisbane, the St. George district, and Normanton in Queensland.

All these places are on the eastern coastal strip among the wooded mountains of the Australian Cordillera and its continuation, the Australian Alps. But the most relevant rumors seemed to be confined to the York Peninsula, at the extreme north of the mountain range.

On August 2, 1871, a police magistrate at Cardwell, Brinsley Sheridan, wrote to the famous zoologist, Sclater:

> One evening strolling along a path close to the shore of Rockingham Bay, a small terrier, my son's companion, took a scent up from a piece of scrub near the beach, and followed, barking furiously, towards the coast-range westwards. My boy (thirteen years of age, but an old bushman, who would put half those described in novels to

the blush) followed and found in the long grass, about half a mile from the spot the scent was first taken up, an animal described by himself as follows:—"It was lying camped in the long grass and was as big as a native Dog; its face was round like that of a Cat, it had a long tail, and its body was striped from the ribs under the belly with yellow and black. My Dog flew at it, but it could throw him. The animal then ran up a leaning tree, and the Dog barked at it. It then got savage and rushed down the tree at the Dog and then at me. I got frightened and came home."

Sheridan made inquiries and found that this was not the first time that such an animal had been seen in the district.

On December 4, 1871, Walter J. Scott, C.M.Z.S., also wrote from Cardwell with recent news of the "tiger." Six men working on the banks of the Murray and Mackay Rivers north of Cardwell were waked up one night in their tent by a "loud roar." They leaped to their guns and cautiously explored the neighborhood. The animal had gone, but around the camp there were tracks of a

39. Drawing of the Queensland marsupial tiger cat based on all the known evidence.

largish carnivore. A picture of a footprint was published with the letter in the *Proceedings of the Zoological Society*. Scott also recalled that in 1864 a bullock driver said that he had met a tiger face to face in the area, but as he was a notorious liar no one believed him. Perhaps, after all, he had seen a similar animal.

Nonsense, said Gerald Krefft, the bullock driver was certainly a liar, and Scott's footprint was an ordinary dog's!

Meanwhile, on June 5, 1872, Scott sent another more detailed report from the Valley of Lagoons in Queensland. Robert Johnstone, an officer of native police, who was in the scrub with several of his men in the coastal area west of Cardwell, saw a large animal perched in a tree some 40 feet from the ground. When they approached it leaped 10 feet into another tree, clung to it, and then slithered down tail first. It was larger than a pointer and its color was fawn with darker patches. Its head was quite round; they did not see its ears. Its tail was long and thick.

In 1895 there was a scare in the press about a "tiger" at Tantanoola in South Australia. The beast was hunted and finally shot. Its dismembered carcass was exhibited for all to wonder at. Alas, it was soon recognized to be a calf's, the show was closed, and those who believed in the "tiger" theory were much disappointed, and consoled themselves by accusing the cowardly hunters of making a scapegoat of a poor calf. Another theory soon prevailed: cattle thieves had kept the "tiger" legend alive so they could steal in peace and blame that bloodthirsty creature for their thefts.

Certainly there was plenty of talk of "tigers" at this time: one was reported at Moruya, another in Gippsland, a third in the forest near Colac, and a fourth in Riverina. When the second was killed it proved to be a pig that had gone wild; the third was merely a wild dog that had escaped from captivity. There was now a tiger complex and people were seeing them everywhere, which naturally did not tend to make young Sheridan's story believed. But a shrewd observer would have remarked that whenever a report came from the north of Queensland it was more positive and more precise.

Thus in 1900, at Kairi, J. McGeehan came upon a "striped marsupial cat" being attacked by a pack of dogs and crying piteously. It was clearly striped in alternate dark brown and white rings about 2½ inches wide. Its head was like a Pomeranian dog's and the length of the whole animal must have been more than 2 feet. No doubt it was a young one.

The animal was also seen by a well-known naturalist, George Sharp, near the source of the Tully River. One evening he heard a rustle in the scrub and came out of his tent. It was not yet dark and he could see an animal "larger and darker than the Tasmanian Tiger, with the stripes showing very distinctly." A little later one of its brethren attacked a farmer's goats on the Atherton Tableland and was killed. Sharp heard of it and at once set off to the place where he was able to examine the animal's skin. It was about 5 feet long from the tip of its nose to the end of its tail. As he had no means of preserving the pelt at hand, it soon decayed. Wild pigs had already eaten the head and body.

A Mr. Endres of Mundubbera in Queensland caught one of these "tigers" alive. He said it was "about 18 inches high and as long as a large cat; very short head and neck; striped, but not right round." It was "very savage when caught."

Ion Idriess, an Australian writer who spent his whole life in the wild northeast of the continent, mainly in Coen, writes of the mysterious creature as if it were a very familiar beast.

> Up here in York Peninsula we have a tiger-cat that stands as high as a hefty, medium-sized dog. His body is lithe and sleek and beautifully striped in black and grey. His pads are armed with lance-like claws of great tearing strength. His ears are sharp and pricked, and his head is shaped like that of a tiger. My introduction to this beauty was one day when I heard a series of snarls from the long buffalo-grass skirting a swamp. On peering through the grass I saw a full-grown kangaroo, backed up against a tree, the flesh of one leg torn clean from the bone. A streak of black and grey shot towards the "roo's" throat, then seemed to twist in the air, and the kangaroo slid to earth with the entrails literally torn out. In my surprise I incautiously rustled the grass, and the great cat ceased the warm feast that he had promptly started upon, stood perfectly still over his victim, and for ten seconds returned me gaze for gaze. Then the skin wrinkled back from the nostrils, white fangs gleamed, and a low growl issued from his throat. I went backwards and lost no time in getting out of the entangling grass.

It was not the only encounter.

> The next brute I saw was dead, and beside him was my much-prized staghound, also dead. This dog had been trained from puppyhood in tackling wild boars, and his strength and courage were known by all the prospectors over the country.

This happened not far from the Alice River.

P. B. Scougall and G. de Tournoeur were riding from Munna Creek to the little town of Tiaro. Suddenly their horses shied.

> We dismounted [they wrote to the *Brisbane Courier*] and were startled to find the cause to be a large animal of the cat tribe, standing about twenty yards away, astride of a very dead calf, glaring defiance at us, and emitting what I can only describe as a growling whine. As far as the gathering darkness and torrential rain allowed us to judge he was nearly the size of a mastiff, of a dirty fawn

colour, with a whitish belly, and broad blackish stripes. The head was round, with rather prominent lynxlike ears, but unlike that feline there were a tail reaching to the ground and large pads. We threw a couple of stones at him, which only made him crouch low, with ears laid flat, and emit a raspy snarl, vividly reminiscent of the African leopard's nocturnal "wood-sawing" cry. Beating an angry tattoo on the grass with his tail, he looked so ugly and ready for a spring that we felt a bit "windy"; but on our making a rush and cracking our stockwhips he bounded away to the bend of the creek, when he turned back and growled at us.

Finally I should mention that the large striped cat has been known by the Queensland aborigines since the earliest times.

It is not therefore surprising that the eminent zoologist E. Le G. Troughton, Curator of the Department of Mammals in the Australian Museum, should write:

> Although there is some divergence concerning the size of the animal and the disposition of the stripes, there seems some possibility that a large striped marsupial-cat haunts the tangled forests of North Queensland.

Nor would there be anything very odd about finding an Australian animal that looked like a cat. That this "tiger" or (more modestly) "tiger cat" is a marsupial is very likely, since it lives in an almost exclusively marsupial world. As I have already remarked, Australia has, by a curious phenomenon of convergence, marsupials that resemble most types of placentary mammals, and in the past it had even more. There was, for instance, until quite recently a sort of marsupial lion, the *Thylacoleo*. It would be quite natural to find a smaller species of the same group on the same continent.

This large striped animal which has been well described as "a cat just growing into a tiger" is, of all the unknown animals dealt with in this book, the one nearest to being officially admitted by science. Although there is neither skeleton nor skin of it in any museum in the world, the "striped marsupial cat" of Northern Queensland was included in a standard work on Australian fauna in 1926 on the strength of the field observations which I have quoted from its text. The authors, A. S. Le Souef and H. Burrell, describe the "striped marsupial cat" of York Peninsula thus:

> Hair short, rather coarse. General colour fawn or grey, with broad black stripes on flanks, not meeting over the back. Head like that of

> a cat; nose more produced. Ears sharp, pricked. Tail well haired, inclined to be tufted at end. Feet large, claws long, sharp. Total length about five feet; height at shoulder eighteen inches.

It would be almost as big as the small clouded leopard or a cheetah with short legs.

So far it seems to be confined to Northern Queensland, where its habitat is "the rough, rocky country on top of the ranges, country usually covered with heavy forest, and inhabited chiefly by tree-kangaroos and rock-wallabies." But Le Souef and Burrell add:

> We have had a striped carnivorous animal described from North-west Australia, and Lord Rothschild states, from native reports, that a similar animal exists also in New Guinea.

And they point out why it is not better known to science.

> This animal is rare, or, to be more correct, it lives in country that man seldom penetrates, and when he does so he creates such a noise in getting through the tangled undergrowth that any wary animal makes off.

There is a rather similar phenomenon in physics. It is theoretically impossible to shine a ray of light onto a particle in order to ascertain its relative position and velocity, without the impact of the light altering them and giving a false result. This impossibility of making extremely small measurements led the German physicist Wernher Heisenberg to formulate his now-famous Principle of Uncertainty in 1925.

There ought to be a similar zoological principle, based on the difficulty of making field observations of large animals, man's natural rivals, which tend to disappear as soon as man arrives on the scene. The degree of difficulty depends on their habitat: obviously it is greater in the water than on land, and greater on land than in the air; greater in mountains than in plains; greater in the forest than in the savanna or the bush, and greater in the bush than in a sandy desert. This is why we know so little about the creatures of the sea and so much about birds—except those species that cannot fly. The difficulties are almost insuperable when two coincide, as they do in mountainous country with thick vegetation. That is why most unknown animals live in marshy or mountainous forests. It is a mere matter of logic.

Bay of Islands
Auckland
T A S M A N
S E A
NORTH
ISLAND
C. Runaway
R. Wharekahike
Bay of Plenty
Hick's Bay
East Cap
L. Rotorua
Taranaki
Rangitaiki
Hikurangi Mt.
R. Waiapu
Whaiti Mts.
Gisborne
Poverty B
L. Taupo
Mt. Egmont
Takaka
Riwaka
Nelson
Cook Str.
Wellington
Cloudy Bay
Buller R.
Marlborough
Paparoa Ra.
Grey R.
Kaikoura
P A C I F I C
Southern Alps
Canterbury
L. Heron
Christchurch
Mt. Cook
Ashburton
SOUTH
ISLAND
Arowhenua
Timaru
Waimate
Secretary I.
L. Te Anau
Thompson Sd
Resolution I.
Dusky Sd
Preservation Inlet
Waikouaiti R.
Dunedin
Otago Peninsula
O C E A N
Molyneux Harbour
Invercargill
Stewart I.
Miles
0
50
100
150
200
Forests

CHAPTER 9

THE MOA, A FOSSIL THAT MAY STILL THRIVE

Of fowles every kinde
That in this world han fetheres and stature
Men myghten in that place assembled fynde . . .
GEOFFREY CHAUCER, *The Parliament of Fowls.*

THE MOA is the oddest of birds. It is, as far as we know, the largest bird that ever lived. Not only is it unable to fly, but it has not the smallest trace of wings. There is not even a sign of the collarbone with which a vertebrate's front legs are almost invariably connected to the thorax. It is the truest of bipeds, and its feathers are so primitive that they seem more like hairs.

But though a wingless furry bird might be a freak elsewhere it would not be out of the ordinary in New Zealand. Not only is there a marked tendency among many different groups of birds on those islands to lose their powers of flight, but New Zealand is also the sole home of two families of wingless and tailless birds with hair-like plumage: the kiwis and the moas. The kiwi has vestiges of wings beneath its feathers. It is a nocturnal forest bird as big as a chicken. Its appearance, a little head and long beak sticking out of a bunch of dark brown silk, is familiar from the many pictures of it as the national emblem of New Zealand and a trade-mark for shoe polish. It is less well known that its nostrils are in the tip of its burrowing beak and it is the only bird that finds its food by smell—indeed the only one that has a highly developed sense of smell. The moas, on the other hand, are completely wingless. Various authors have described from 21 to 38 species belonging to five distinct genera, though they are all—officially at least—said to be fossils.

It is usual to talk of kiwis as dwarfs and moas as giants. The truth is less simple. The largest moas, such as the *Dinornis robus-*

40. The New Zealand kiwi.

tus, stood nearly 12 feet high—3 feet higher than the tallest ostrich—but some, like the *Anomalopteryx parva,* were no bigger than a turkey. On the other hand, it is generally admitted that not long ago there was—and perhaps still is—a giant Apteryx (*Apteryx maxima*) which was just the same size.

Why, you may ask, is it in New Zealand rather than elsewhere that birds tend to lose the use of their wings, and even their wings themselves? The answer is to be found in a fact which Captain Cook noticed when he explored the New Zealand archipelago in 1773, and found no quadrupeds except dogs and rats.

These were obviously introduced by the natives, so there were no indigenous quadrupeds in this country. Thus there were no carnivores that eat birds.

In islands where there are no predatory mammals, birds no longer need to take to the air at the least danger, but can relapse into a lazy terrestrial existence—with fatal results. It is hardly surprising that most birds that have recently become extinct have lived on islands, inaccessible to carnivores, where they have given up flying. For sooner or later man and his parasites, the dog and sometimes the cat, come to disturb the peace. The most celebrated victim is the dodo of Mauritius (*Didus ineptus*), a huge flightless pigeon which man exterminated in less than a century. Its name is a very symbol of extinction: as dead as the dodo.

Of all the islands of any size, the most isolated from any mainland are the North and South Islands of New Zealand. They are more than 1,000 miles from Australia, the nearest land. Before the end of the Cretaceous period, these islands' link with the rest of the world sank under the waves forever. This happened shortly before the great invasion of marsupials, which were thus unable to spread beyond Australia. Since then the only access to New Zealand has been by air or by sea. There the only known indigenous mammals are two species of bats, both called *pekapeka* by the Maoris. When land mammals first set foot on New Zealand it was because they had deliberately steered the tree trunks there which they had hollowed out into boats, for they were men. They brought with them their faithful companion the dog, *kararehe,* now vanished. The rat *kiore* (*Mus exulans*) had certainly not been invited, but its arrival was no accident.

During the long period between New Zealand's isolation and its invasion by man, birds were able to develop richly and into the wildest forms, such as the giant moas. After losing their power of flight, the birds, isolated from one another in the two main islands, evolved along different lines, as is explained by Von Hochstetter.

> According to Prof. Owen, the birds of South Island present stouter proportions, a compact, rather bulky frame of body, such as *Dinornis robustus, elephantopus, crassus* and *Palapteryx ingens,* while those of North Island are distinguished by more slender and lengthy forms, like the *Dinornis giganteus* and *gracilis.*
>
> These various species inhabited the plains and valleys and had their hiding-places in forests and caves. Their food doubtless consisted of vegetables, especially fern-roots, which they dug up with their powerful feet and claws. To assist the process of digestion, they swallowed small pebbles.

This we know because heaps of small round stones, usually chalcedony, cornelian, agate, and opal, are often found with the bones of moas or occasionally by themselves. Presumably these round pebbles came from the moas' stomachs, for their cousins the ostrich and the emu are in the habit of swallowing small pebbles to help grind the food and of regurgitating them from time to time when they have been worn quite smooth.

> The formation of the skull [Professor Von Hochstetter goes on] leads us to infer that they were stupid, clumsy birds, which we must not suppose to have been swift runners like the ostrich, but sluggish diggers of the ground, the nature and habits of which demanded no larger scope than such as the limited territory of New Zealand presented.

41. The conventional reconstruction of a large moa,

Such birds would be most vulnerable as soon as the first enemy arrived. But did man exterminate the moa? Yes, say some, going so far as to cite the evidence of Maoris who point out a Totara tree on Lake Rotorua as the place where their ancestors killed the last moa.

> But for those colossal birds [says Von Hochstetter], it would be indeed utterly impossible to comprehend how 200,000 or 300,000 human beings could have lived in New Zealand, a country which even in its vegetable world offered nothing for subsistence, except fern-roots.

Others no less hotly deny that man exterminated the moa, but this can have two opposite meanings. Some say that the moa could not have been exterminated by man, because it was already extinct when he arrived. The degeneration of this overgrown bird, reflected in its bones, made the moa die out. But others say that the moa has not been exterminated by man because it is not yet utterly extinct, and this giant bird, or at all events *some* moa of smaller size, survives in unexplored regions of New Zealand. The extraordinary diversity of opinion on this subject is largely a result of the bitter controversy aroused by the original discovery of the moa.

The history of the discovery of the moa is even more confused than the history of the moa itself. When the moa was discovered it was variously pronounced to be a contemporary of present-day man, a bird only recently extinct, and a complete fossil. Each of these opinions had its own belligerent faction which became ever firmer in its own conviction, after searching systematically for arguments to support its theory and neglecting any evidence that might weaken it. Let us try to follow the moa's discovery without any such bias.

It officially entered the world of science in 1839 when a fragment of a huge bone was brought back from New Zealand and eventually reached the hands of Professor Sir Richard Owen. On November 12, 1839, he read a note on this sensational discovery to the Zoological Society of London in which he said:

> I am willing to risk the reputation . . . on the statement that there has existed, if there does not now exist, in New Zealand, a Struthious bird nearly, if not quite, equal in size to the Ostrich.

There was no doubt that such a bird had once existed when the bone had been identified by Owen. But that it could still exist or be known to man seemed unlikely. Captain Cook had heard nothing of it when he visited New Zealand in 1773. And no traveler had made the least mention of this extraordinary bird—at least until 1838. Then J. S. Polack, a Jewish merchant, published two volumes about New Zealand, having spent more than two years on the east coast of North Island, where, he said:

> That a species of the emu, or a bird of the genus Struthio [ostrich], formerly existed in the latter island, I feel well assured, as several large fossil ossifications were shewn to me when I was residing in the vicinity of the East Cape, said to have been found at the base of the inland mountain of Ikorangi [Hikurangi]. The Natives

added that, in times long past, they received the traditions, that very large birds had existed, but the scarcity of animal food, as well as the easy method of entrapping them, has caused their extermination.

This would have satisfied those who thought the moa was recently extinct and explained why Captain Cook had heard nothing of such a monstrous bird being still alive. But Polack goes on to say:

> I feel assured, from the many reports I received from the natives that a species of struthio still exists on that interesting island, in parts, which, perhaps, have never yet been trodden by man. Traditions are current among the elder natives of Atuas,* covered with hair, in the form of birds . . .

Although a New Zealand printer and amateur naturalist called William Colenso was later to maintain that Polack did not even know how to write—so that he could pretend that he himself was the first to mention the moa—the book is still the first record of the giant bird. Colenso can, however, claim to be the first to mention the name *moa* in print. He writes:

> During the summer of 1838, I accompanied the Rev. W. Williams on a visit to the tribes inhabiting the East Cape District. Whilst at Waiapu, (a thickly inhabited locality about 20 miles S.W. from the East Cape) I heard from the natives of a certain monstrous animal; while some said it was a bird, and others "a person," all agreed that it was called a *Moa;*—that in general appearance it somewhat resembled an immense domestic cock, with the difference, however, of its having a "face like a man";—that it dwelt in a cavern in the precipitous side of a mountain;—that it lived on air;—and that it was attended, or guarded, by two immense *Tuataras* [*Sphenodon*], who, Argus-like, kept incessant watch while the *Moa* slept; also that if any one ventured to approach the dwelling of this wonderful creature, he would be invariably trampled on and killed by it.

On the strength of this Colenso pretended that he was the first white man to learn that a giant bird existed in New Zealand. But it seems clear to me that he did not believe a word of these fantastic legends, or he would have discussed them with Williams. For when this missionary returned to the east coast a year later in January 1839 with the Reverend Richard Taylor, he was skeptical, or perhaps even ignorant of any such legends. Taylor writes:

* The Maori name for all supernatural beings from the supreme being down to mere demons.

> In the beginning of 1839 I took my first journey in New Zealand to Poverty Bay with the Rev. Wm. Williams (the present Bishop of Waiapu). When we reached Waiapu, a large pa near the East Cape, we took up our abode in a native house, and there I noticed the fragment of a large bone stuck in the ceiling. I took it down, supposing at first that it was human, but when I saw its cancellated structure I handed it over to my companion, who had been brought up in the medical profession, asking him if he did not think it was a bird's bone. He laughed at the idea, and said, what kind of bird could there be to have so large a bone? I pointed out its structure, and when the natives came requested him to ask them what it belonged to. They said it was a bone of the Tarepo, a very large bird which lived on the top of Hikurangi . . . in a cave, and was guarded by a large lizard, and that the bird was always standing on one leg.

Taylor sent the bone to England, where it eventually reached Owen's hands but not until he had already seen the bone I mentioned earlier. So Taylor's claim is as unjustified as Colenso's when he writes, "I first discovered its remains in 1839, at Tauranga [Gisborne] and Waiapu."

When Taylor returned to the Bay of Islands in February he met another person who claimed to have heard of the moa, the German naturalist Dr. E. Dieffenbach. When he tried to climb Mount Egmont, a native chief dissuaded him because the mountain was guarded by a moa or *movie.*

Colenso was then a neighbor of Dieffenbach at the Bay of Islands, so he probably learned the German traveler's story and was reminded of the fantastic tales he had heard earlier about a devil bird with a human face. When, like Dieffenbach, he had an opportunity of examining the bones that Taylor had brought back—in particular a toe—he must have realized for the first time that the native legends were not utterly without foundation.

Two years later Colenso decided to find out more about the mysterious bird. At Rangitukia, on the left bank of the River Waiapu, he obtained his first moa bones: five femurs, a tibia, and one unidentified bone. At Poverty Bay he met his old friend the Reverend William Williams, who had also collected a number of moa bones from the natives, intending to send them to Professor Buckland at Oxford. Colenso went on scouring the east coast, asking the Maoris everywhere whether they knew of moa bones. They showed him a footprint in a clayey rock, a little beyond Hicks Bay, and said it was that of Rongokako, one of their famous ancestors. The next footprint was near Poverty Bay, more than 80 miles away—outdoing even the seven-league boots of fable! Colenso did not realize that it was actually a moa's footprint.

When he returned to the Bay of Islands in April 1842 he feverishly set about completing his collection of moa bones so that he could send them posthaste to his friend Sir William J. Hooker for Sir Richard Owen. Determined at all costs to be the first to have told the world of the giant fossil bird, Colenso always swore that he had never seen Owen's 1839 note, nor to have heard of it or have met any settler that knew of it. In fact the New Zealand papers, as one might expect, were full of echoes of this local wonder after Owen's note appeared, and more than a hundred offprints of the note itself were distributed in New Zealand. The missionaries, in whose wake Colenso followed, were the most educated of the settlers and would have certainly been the first to receive copies.

Many people in New Zealand claimed to be the first: the first to hear legends of the giant bird, the first to hear the native name of *moa,* the first to collect its bones. And as they often distorted the truth and contradicted themselves in order to establish their claim, the first reports have often seemed suspect since. They seemed even more suspect when there were obstinate rumors that the moa survived. It was no longer possible to pretend to be the first to have seen bones or heard the name, so perhaps people were looking for another claim to fame: to be the first to see a living moa. The seed of skepticism was sown.

In 1842 Archdeacon Williams sent a vast quantity of moa bones packed in huge wicker baskets to Dr. Buckland, who gave them to the Royal College of Surgeons. From these bones Sir Richard Owen was able to reconstruct the enormous legs of *Dinornis giganteus,* which must, by his reckoning, have stood more than 10 feet high. He also established that there were at least four different species. These were the conclusions of his second note published in January 1843.

Then moa bones were found in South Island for the first time, at the mouth of the Waikouaiti by Percy Earle and Dr. MacKellar. But the richest harvest of all was undoubtedly that which Walter Mantell collected on both islands between 1847 and 1850. It included no less than 1,000 separate bones and even pieces of the shells of huge eggs that would have been as much as 10 inches long. These completed the rich material upon which Sir Richard Owen based his now famous monograph *On Dinornis.* Among them was the skeleton of a moa hardly 5 feet high, but so heavy and massive that Owen called it *Dinornis elephantopus* ("the terrible bird with elephant's feet"). Thus there were not only versions of attenuated giraffes and small turkeys among the moas, but even of pachyderms.

But when did these legendary birds live? Some of the disinterred bones are unquestionably fossilized and can be dated from the middle of the Miocene period some 15,000,000 to 35,000,000 years ago. But most moa bones are in a *subfossil* state. That is to say they have not had time to "petrify," or be impregnated with mineral salts, and Hochstetter says that their age can only be counted by hundreds, instead of thousands of years.

As the Maoris had hardly settled in New Zealand two or three centuries before it was sighted by Tasman in 1642, the relative freshness of the bones is not enough to prove for certain whether man ever knew the living moa. We must therefore turn to the local traditions.

The legends that Colenso collected hardly indicate an intimate knowledge of the living bird. Its description as a sort of giant cock with a man's face seems very fanciful. And the fact that the Maoris attributed its footprint to legendary heroes or demigods seems to prove that they could not have known it. Hill therefore concluded in 1914 that "not a single Maori known to the missionaries and early settlers ever saw or heard of a moa, or of its having been seen by any of their ancestors." Everything that the natives said about the moa was a myth based on the discovery of huge bones and on what they had learned from white men. Hill maintained that the moa died out in the Pleistocene period, and probably at the beginning of it, between 300,000 and a million years ago, but this is contrary to the paleontological evidence and betrays surprising ignorance of some aspects of the problem. All the first reports about the moa—upon which Hill based his arguments—came from North Island, and especially from the east coast. It seems likely that the natives in this area did not know the moa directly. But how could they have known that the huge bones belonged to a bird? They were not paleontologists. One might argue that since almost all the other animals on the islands were birds they would have assumed it to be a bird, and the remains of its eggs would have confirmed the theory. But how did they know what its plumage was like? Polack refers to "Atuas, covered with hair, in the form of birds." Of all the New Zealand birds only the kiwi has feathers like silky hair, and the natives could not have guessed that the bones and eggs belonged to an outsize kiwi and not to some other type of bird.

So if the Maoris of the east coast of North Island did not know the moa themselves or from their immediate ancestors, their legends are still originally based on an exact description of the bird. Other men must therefore have known it, travelers, perhaps, from more southerly corners of New Zealand. And, in 1844, when Governor Fitzroy met the Maoris at Wellington at the very south of North Island, Haumatangi, an old Maori of 85 who boasted that he had seen Captain Cook in 1773 when he was only 14, said that he had seen a moa for the last time two years before that memorable date. Kawana Papai, another no less ancient native, maintained that he had himself taken part in moa hunts on the Waimate plains in South Island some 50 years before. The birds were rounded up, encircled, and then killed with spears—not an easy task or a safe one, since the moa, like the ostrich, fought hard, laying about it with terrible blows of its feet. The moa hunters used special spears which were designed to break off inside the bird's body. Once dead, it was cut up with an obsidian knife, and eaten at gargantuan feasts. On both North Island and South Island, Mantell and another naturalist called Cormack often found mounds full of charred moa bones near native camps and settlements, the remains, no doubt, of these meals. Thus, for the natives of South Island and of the south of North Island, the moa, which might weigh as much as a quarter of a ton, was as welcome in the larder as a Christmas turkey.

On South Island, not only have fresher remains of moas been found, mummified pieces of the body, the skin and brownish feathers with white tips still preserved, but much more widespread and detailed native traditions have been collected. Von Hochstetter remarks that there are Maori poems in which the father tells his son exactly how to hunt the moa. He also cites from Maori tales the use that was made of the moa:

> The flesh and eggs were eaten; the feathers were employed as ornament for the hair; the skulls were used for holding tattooing powder; the bones were converted into fish-hooks, and the colossal eggs were buried with the dead as provision during their long last journey to the lower regions.

The Maoris said it had a brilliant plumage and a high crest on its head. Altogether it looked like a huge Cochin cock, a fat, gallinaceous bird with feathered legs. This apparently absurd description has been used as evidence that the Maoris had never seen moas. Actually it is our reconstruction that is more likely to be absurd. It is largely based on the emu, which has bare legs and sober grayish-brown plumage. But we have no proof that *some* moas did not have brilliant plumage like that other Australian bird, the cassowary. The brown feathers that have been found may belong to females, usually duller in color than the males. The

42. Probable appearance of some moas, according to the Maoris who knew them.

moa's neck could have been as thickly feathered as an emu's, which would have completely altered its outline. And it could have had feathered legs as the kiwi does. Only a fast-running bird of the plains has any advantage in bare legs, and the moa was not one.

Thus it seems that innumerable moas still lived in New Zealand when it was first invaded by man. He came from the north, and both islands must have been invaded from the north or northwest and the moa driven farther and farther to the south. On both islands moa bones have been found in the largest quantities in the southeast. There is little doubt that most of the moas on North Island had vanished 500 years before man arrived on the island. Perhaps a few isolated flocks of birds were killed as late as 1800 in the extreme south of the island, as the remains of feasts indicate, but almost all of the accounts reported by Governor Fitzroy were either based on tradition or came from Maoris on the other side of Cook Strait. In South Island the moa—particularly *Euryapteryx gravis*—was not only well known to the Maoris at one time, but played such an important part in their lives that Roger Duff could write a learned work entitled *The Moa-Hunter Period of Maori Culture* (1950). According to him this period ended in the sixteenth century. But there is also proof that in South Island the moa survived until very recently and certainly existed when New Zealand was discovered by white men. All that remains is to discover just when the last of the moas died.

Colenso and Taylor thought that the moa survived into the nineteenth century. And in 1863 Von Hochstetter wrote: "It is by no means impossible that in lonely and inaccessible places live a few scarce diehards of the giant family: the last of the Mohicans." At the beginning of the nineteenth century American sailors and sealers claimed to have seen monstrous birds 12 to 15 feet high (an exaggeration, no doubt) running to and fro along the coast of Cloudy Bay, on South Island, and on the inhospitable shores of the southwest of that island. According to Taylor a sealer called Meurant found moa bones still covered with flesh in 1823 at Molyneux Harbour on South Island. Finally there is this report quoted from the *Nelson Examiner* of January 12, 1861, by Von Hochstetter:

> In June, while Messrs. Brunner and Maling, of the Survey Office, were surveying on the ranges between Riwaka and Takaka, they observed one morning the foot-prints of what appeared to be a very large bird, whose track, however, was lost among the scrub and rocks. The foot-prints were 14 inches in length, with a spread of 11 inches at the points of the three toes. Similar foot-prints were seen on a subsequent morning, and as the country is full of limestone-caves, it is thought that a solitary Moa may yet be in existence.

Von Hochstetter did not believe all these stories. According to him the largest living running bird in New Zealand was certainly a giant Apteryx (*Apteryx maxima*), although actually no more is known about this bird, which is included in the present fauna of New Zealand, than about the supposedly extinct moas.

Three species of kiwis are known today: *Apteryx mantelli* confined to North Island, *Apteryx australis* confined to South Island, and *Apteryx oweni* found in both. Where two kinds are found together the Maoris differentiate them by their size, calling the larger *kiwi-nui* and the smaller *kiwi-iti*. None of them is larger than a domestic fowl. But there was once a much bigger species called *roa-roa*. Von Hochstetter quotes two reports of "a Kiwi about the size of a turkey, very powerful, having spurs on his feet, which, when attacked by a dog, defends himself so well as frequently to come off victorious."

I do not believe that any bones of this mysterious large kiwi have ever been found. There are only a few scraps of skin covered with feathers which adorn a Maori chief's ceremonial cloak. So why have some authorities considered it as a member of the genus *Apteryx?* Kiwis have no "spurs" or well-developed hind claws. Would it not therefore be more likely to be a small moa, an *Anomalopteryx* for instance, which was exactly the size of a turkey?

Sir George Grey, Prime Minister of New Zealand, told Sir Walter Buller that in 1868 he was at Preservation Inlet and saw a party of natives there who gave him "a circumstantial account of the recent killing of a small Moa (?*Palapteryx*), describing with much spirit its capture out of a drove of six or seven."

Much larger moas may have survived until today. We know from Maori evidence that they were shy birds that liked to be alone. It has always been supposed that the Dinornithidae, like other running birds, preferred to live in grassy plains. Actually they were a very varied family, large and small, slender and squat. It is very likely that the different kinds had no less different habitats, and that there were slim moas of the plains, little moas of the forests, and heavy moas of the marshes—like the aepyornis of Madagascar. And some of them may well have taken refuge, like the kiwi, in nocturnal habits. Nocturnal moas, living in forests, would have a good chance of eluding all our searches.

That New Zealand can still hide an unknown animal of fair size is strikingly proved by the story of another odd bird, the *takahe*.

When the moa hunt first began in North Island, the Maoris often mentioned another bird, the *moho* that they had formerly eaten, but it had now vanished. It was as big as a goose, and it could not fly although it had wings and feathers that looked like real feathers. But in all the harvest of moa bones there was none of the *moho,* and the bird was shelved with other Maori myths until 1847, when Walter Mantell obtained a skull and several bones which he sent to London, saying that they must have come from the bird known as *moho* in North Island and *takahe* in South Island. Owen confirmed the truth of the native reports and gave the bird the name of *Notornis mantelli.*

In 1849 sealers pitched their camp on Resolution Island, one of the largest islets at the jagged southwest tip of South Island. Their dogs led them on the trail of a large bird in the snow until they saw it furiously screaming and flapping in a dog's jaws. It was beautifully colored with dark blue and violet on the head and neck, olive spotted with brighter green on the back and wings, metallic blue on the primaries and rectrices, purplish-blue on the chest and sides, and white under the rump. The thick beak and solid legs and feet were brilliant red.

43. The *takahe* of South Island, New Zealand.

The sealers naturally put the bird in the pot, but they kept the skin in the hope of getting some money for it. By an extraordinary stroke of luck it was bought by Walter Mantell, who was able to let them know in London that the bird that had vanished from North Island still lived on the South. In 1850 a Maori caught a second *takahe* on Secretary Island, not far away from the first. It was also obtained by Mantell, who sent its feathered skin home to join the other in the Natural History Museum.

Then the bird vanished for thirty years. When Von Hochstetter landed in New Zealand in 1858 he tried hard to come by a *takahe* but on his return to Europe he was obliged to write, "it appears to me, that this family of birds is now totally extinct."

No one had any luck, although a Maori of the fiords said that the bird was not at all rare on the shores of Lake Te Anau. In 1861–62 a Dr. Hector found *takahe* tracks near Thompson Sound and by the central part of Lake Te Anau; in 1866 a botanist called Gibson actually saw one in tall marsh grass near Motupipi; and in the same year the Maoris who told Sir George Grey about hunting small moas insisted that the *takahe* was very common. But nothing more happened. Scientists in Europe grew tired of waiting and wrote it off as extinct.

And extinct it would have remained had not a dog come to the rescue again. This dog was out rabbiting in the Province of Otago near the fiords in December 1879 and brought back a *takahe* still breathing to its master, who dispatched the bird and hung it from the ridgepole of his tent, intending to eat it. Fortunately the Station Manager, Connor, happened to pass the next day, and recognized it. He carefully prepared the skin and skeleton and sent it to London, where it was bought by the Dresden Museum. A detailed examination showed that it differed noticeably from that of the *moho* found at Waingongoro. The new species was called *Notornis hochstetteri* as a consolation prize for the scientist who had searched so hard to find it, and with so little luck.

Then the *takahe* played hide-and-seek for another 20 years. Nothing was found except an incomplete skeleton in 1884 and a complete one in 1892—both near Lake Te Anau. The next live *takahe,* in 1898, was once again caught by a dog, whose master recognized the bird and preserved it entire—viscera and all. This specimen, the only one in the world, was sold for £250 to the New Zealand Government.

Then half a century went by without a sign of the *takahe*. There were still vague rumors about footprints in the snow and reports from old Maoris who said that the birds lived in the mountains and only rarely came down near the lakes. But these tales were of just the sort that most zoologists obstinately refuse to notice. This book is full of animals known only from their footprints and from native legends: evidence that is always greeted with amused disbelief. Every time that this bird managed to avoid its hunters—and their dogs—for a decade the authorities decided that it must be extinct.

By 1947, just a century after Walter Mantell had proved the existence—or former existence—of *Notornis,* only four living specimens of the bird had been found and there seemed little hope of seeing a *takahe* again. But Geoffrey Orbell, a doctor at Invercargill, was undaunted, and optimistically set off with several companions into the thick forests on the eastern shore of Lake Te Anau. Three thousand feet up in the mountains he was surprised to find a small lake which scouts had told him of, although it did not appear on any map. This lake was well known in Maori tradition as *Kohaka-takahea*—"the nesting-place of the *takahe*." They heard unusual bird cries and found fresh footprints that might have been made by their quarry. But no *takahe*. Dr. Orbell was not discouraged. He returned to the place in November 1948, loaded with still cameras and motion-picture cameras. His optimism was rewarded. Two *takahes* were caught in a single snare. They were tethered by one leg, photographed from every possible angle, and then released.

A year later a dozen adults were seen and the population of two colonies in neighboring valleys was estimated at between 50 and 100.

Dr. Orbell, encouraged by his success, has set about trying to capture a living *Megalapteryx*. The skeptics will object that the *takahe* never vanished for very long, but that the moa has been lost for a century or two. The recent resurrection of the *cahow* should answer this objection.

Until lately it was universally thought that the Bermuda petrel (*Pterodroma cahow*) had disappeared between 1609 and 1621, exterminated by settlers hungry for its flesh and eggs. The annals of the period report that some pioneers sent to Cooper Island, at the eastern end of the Bermudas, stuffed themselves with *cahows* until they died of indigestion. This orgy, which the famine in the Bermudas helps to explain, was equally fatal for the poor birds, who seemed to have been exterminated.

But in 1951 a team of American naturalists working with Dr. Robert Cushman Murphy were staggered to come upon *cahow* nests, apparently in use, on the cliffs of small islands off Castle Harbour. Five live birds were caught, examined, ringed, and released. Seventeen nests were counted.

The Bermuda petrel had been thought to be extinct for more than three centuries, and it is a flying bird, far more conspicuous and visible than any flightless bird, however large. Today it is resurrected, and one day the moa may well be resurrected too.

Plates 1 and 2. The Andean wolf (*left*), compared with the wolf of the pampas (*right*), was described from a single skin in 1949.

Plate 3. The hoatzin of the Amazon, whose chicks still have claws on their wings.

Plate 4. Fossil imprint of the Archaeopteryx, the reptile bird of the Jurassic period, which had claws on its wings and also teeth.

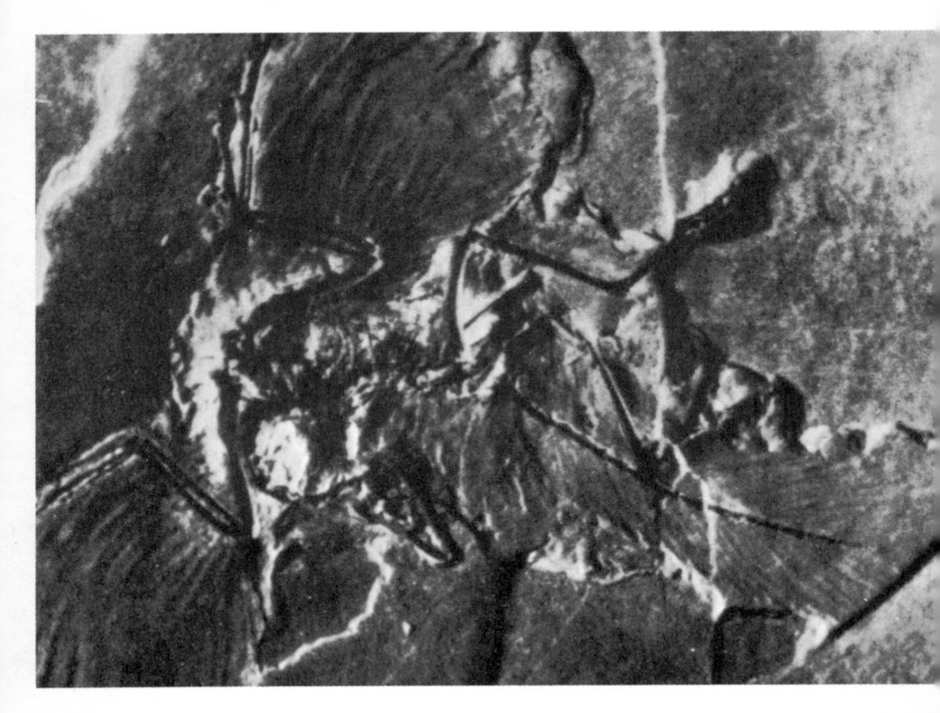

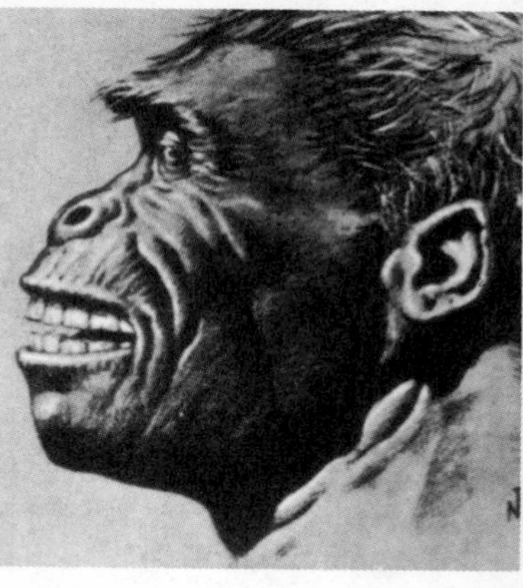

Plate. 6. Reconstruction of the Swartkrans apeman drawn by Neave Parker on evidence by anthropological experts.

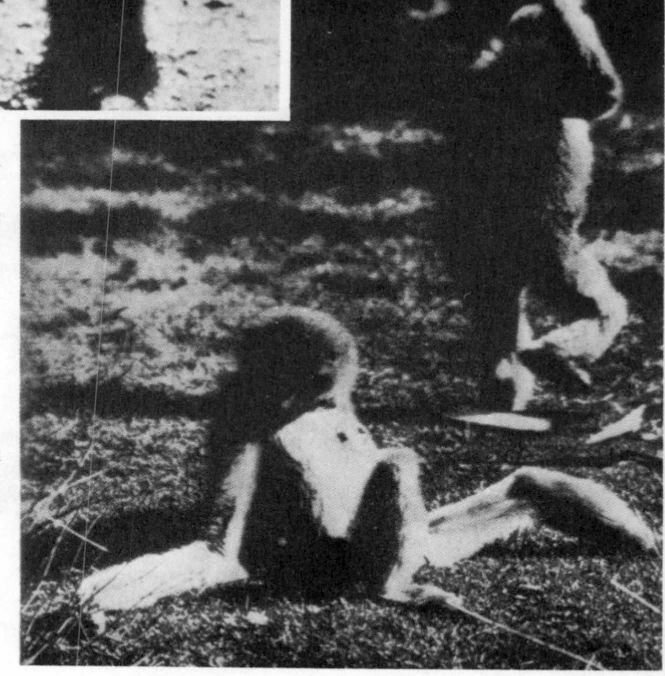

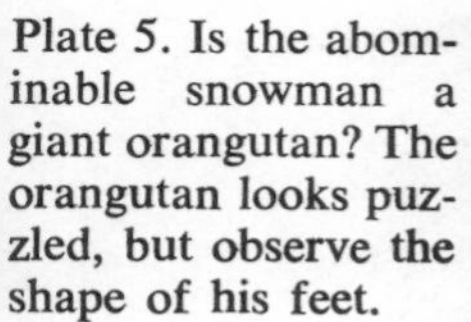

Plate 5. Is the abominable snowman a giant orangutan? The orangutan looks puzzled, but observe the shape of his feet.

Plate 7. The gibbon (*right*) is the only ape that is a biped by nature.

Plate 8. The abominable snowman's trail, photographed by Eric Shipton 1951 with (*inset*) a single footprint compared to an ice-ax. Note how footprints alternate regularly to left and right.

Plate 9. The brown bear's trail photographed on the Col de la Gentiane, Basses-Pyrenées, by Dr. Marcel Couturier. Note how the footprints are often superimposed and blurred, and the claw-marks visible.

Plate 10. The head lama of Khumjun holding a snowman scalp.

Plate 11. The Pang boche scalp, view of the side and th inside.

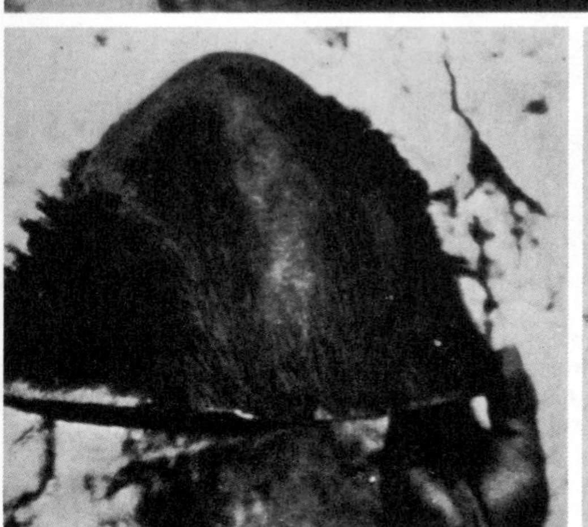

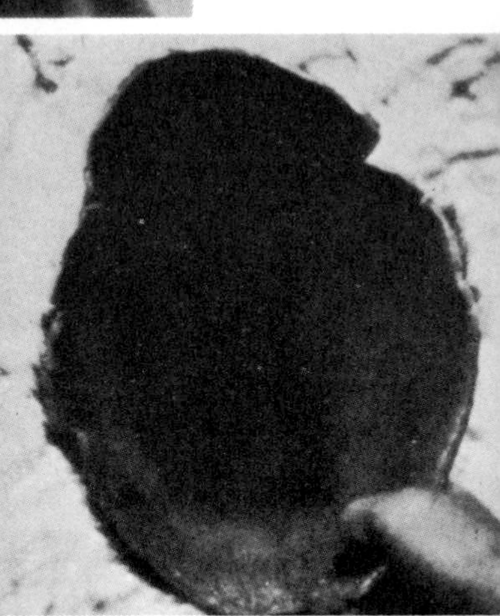

Plate 12. *Ameranthropoides loysi,* the only "unknown" animal of which there is a good photograph.

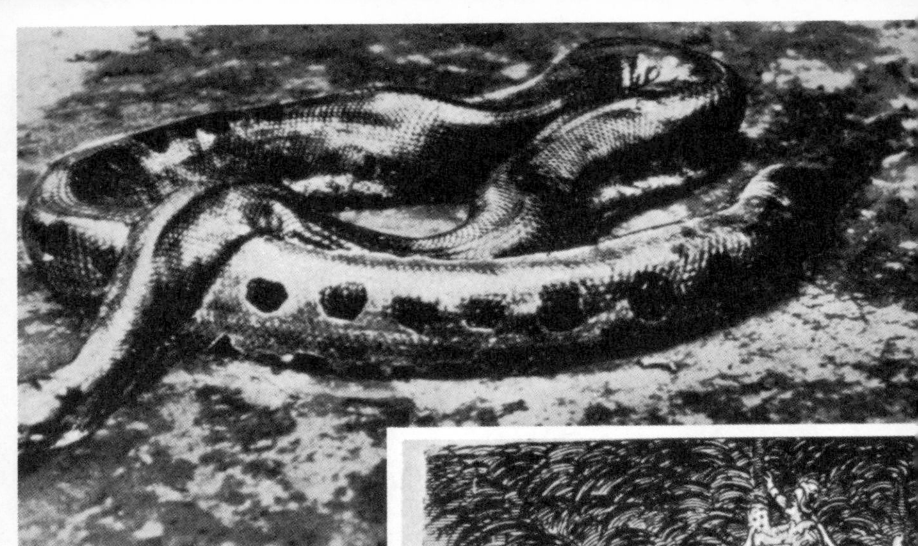

Plate 13. The anaconda is never officially more than 20 feet long.

Plate 14. But there have always been tales of much larger snakes, some of them more plausible than this frontispiece of John Browne's *Affecting Narrative.*

Plates 15 and 16. The spider monkey is undoubtedly like *Ameranthropoides loysi,* but it is much smaller and has a long prehensile tail.

Plate 17. The Berezovka mammoth reconstructed in the Leningrad Museum as it was found, with the trunk partly eaten by wolves. The fur is original

Plate 18. Mammoths' tusks brought from the Liakhov islands to a trading post in Yakutsk. This trade goes back to the earliest times.

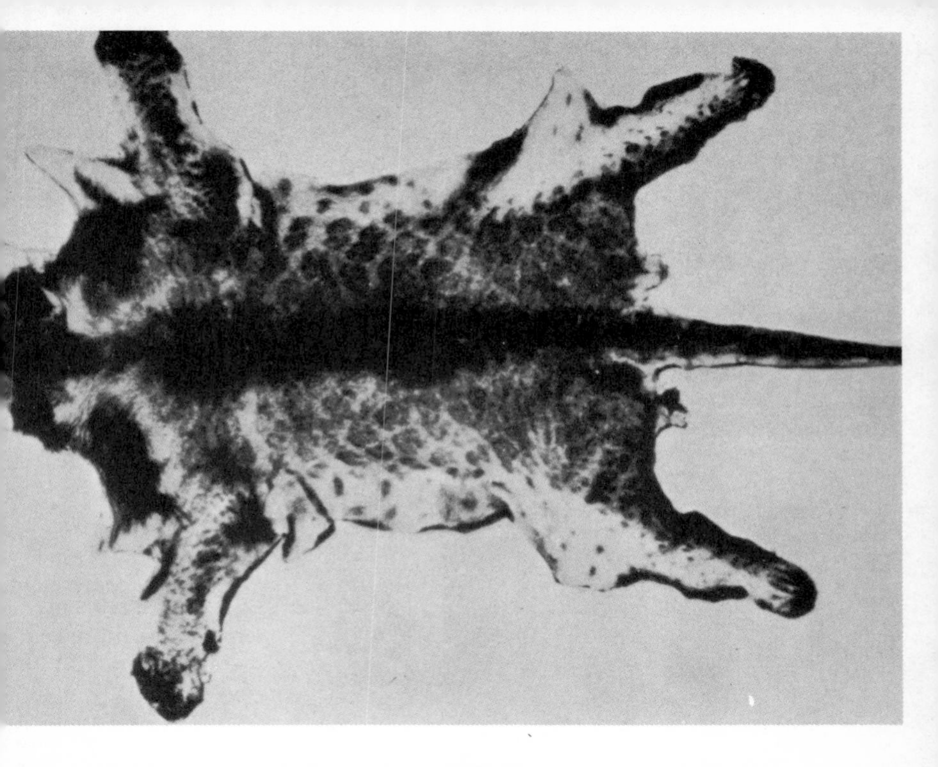

Plate 19. Skin of the spotted mountain lion shot by Michael Trent.

Plate 20. Lion cub showing the spots on its juvenile coat which later vanish.

Plate 21. The spotted h
ena has been blamed
the Nandi bear's crim

Plate 22. The ratel (
low) which becomes qu
black with age, has c
tainly been mistaken
the Nandi bear at tim
Moreover it is a sav
and aggressive beast.

Plate 23. The mandrill may look as savage as a lion. A giant species might be the origin of the Nandi bear.

Plate 24. Modern reconstruction of the Ceratosaurus by Charles R. Knight.

Plate 25. The *sirrush* or dragon of the Ishtar Gate.

Plate 26. East face of the Ishtar Gate showing alternating dragons and oxen.

Plates 27–29. The lemurs of Madagascar which until recently had much larger relatives. (*Above*) *Maki gidro*. (*Left*) *Maki vari*. (*Below*) *Macaco,* an example of sexual dimorphism rare in mammals.

PART FOUR

RIDDLES OF THE GREEN CONTINENT

"Now, down here in the Matto Grosso"—he swept his cigar over a part of the map—"or up in this corner where three countries meet, nothin' would surprise. . . ."

SIR ARTHUR CONAN DOYLE, *The Lost World*.

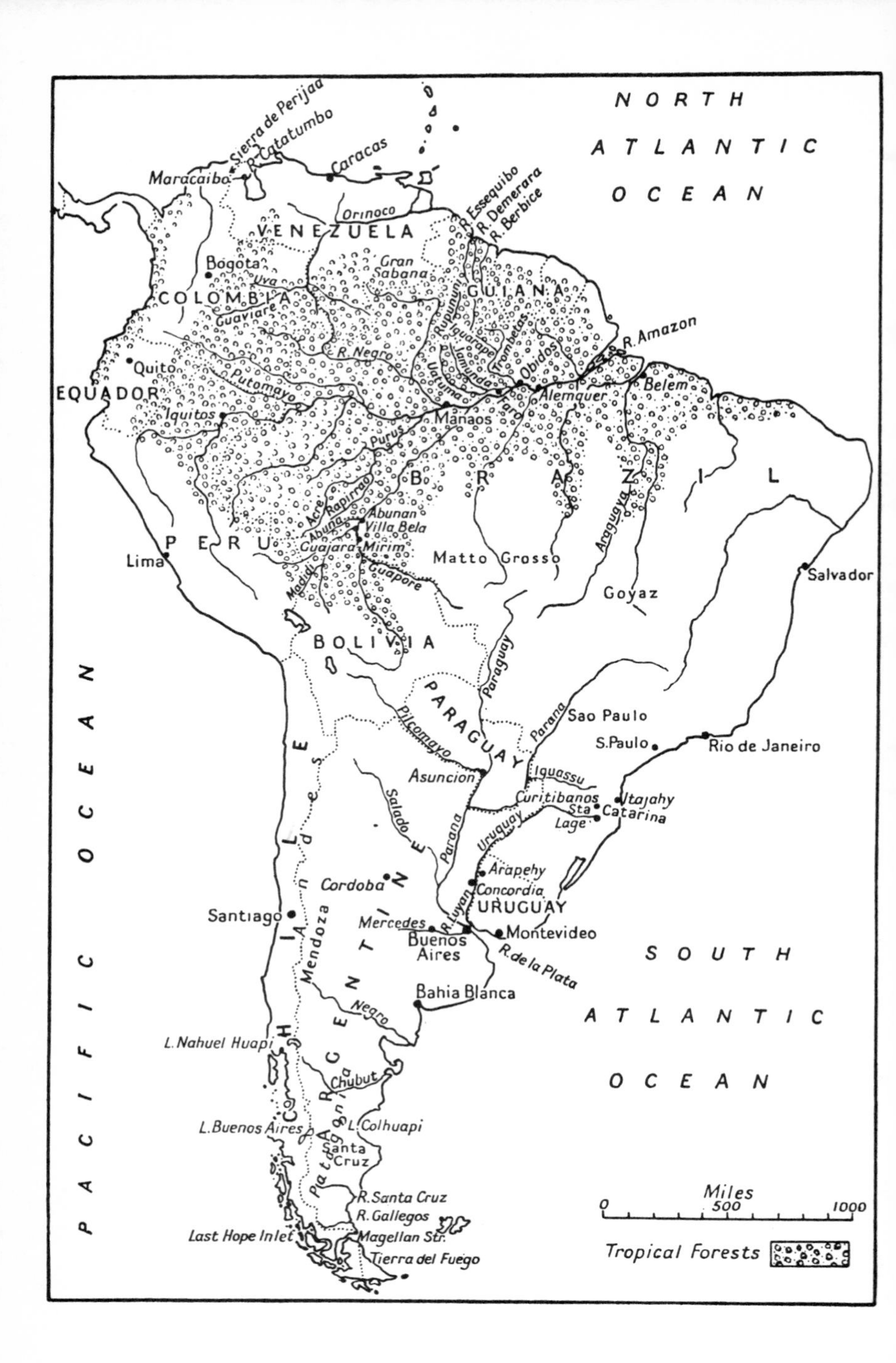
NORTH ATLANTIC OCEAN
SOUTH ATLANTIC OCEAN
PACIFIC OCEAN
VENEZUELA
COLOMBIA
EQUADOR
PERU
BOLIVIA
BRAZIL
GUIANA
PARAGUAY
URUGUAY
ARGENTINE
CHILE
Andes
Sierra de Perijaa
R. Catatumbo
Maracaibo
Caracas
Orinoco
R. Essequibo
R. Demerara
R. Berbice
Bogota
Uva
Guaviare
Gran Sabana
Rupununi
Iguarape
Jamunda
Trombetas
Obidos
R. Amazon
R. Negro
Quito
Putomayo
Iquitos
Uaruma
Manaos
Tapajos
Alemquer
Belem
Purus
Acre
Rapirrao
Abuna
Abunan
Villa Bela
Cuajara Mirim
Madidi
Guapore
Matto Grosso
Araguaya
Goyaz
Salvador
Lima
Paraguay
Pilcomayo
Parana
Sao Paulo
S.Paulo
Rio de Janeiro
Asuncion
Iguassu
Curitibanos
Itajahy
Sta Catarina
Lage
Salado
Uruguay
Arapehy
Concordia
Cordoba
Santiago
Mendoza
R. Luyan
Mercedes
Montevideo
Buenos Aires
R. de la Plata
Bahia Blanca
Negro
L. Nahuel Huapi
Chubut
L. Buenos Aires
L. Colhuapi
Santa Cruz
Patagonia
R. Santa Cruz
R. Gallegos
Last Hope Inlet
Magellan Str.
Tierra del Fuego
Miles
0
500
1000
Tropical Forests

CHAPTER **10**

THE PATAGONIAN GIANT SLOTH

Only one fault he had, which cock-robins have likewise . . . that when any one else found a curious worm, he would hop round them, and peck them, and set up his tail, and bristle up his feathers, just as a cock-robin would; and declare that he found the worm first; and that it was his worm: and, if not, that then it was not a worm at all.

CHARLES KINGSLEY, *The Water-Babies.*

THE CONQUISTADORS who first discovered Central and South America must have been disappointed to find that although the fauna was so rich and new, there were no monsters. From the beginning, naturalists have seen the so-called Neotropical zone, which stretches from Mexico and the West Indies to the very tip of South America, as a region of small animals. In fact, the situation is slightly more complicated. The types of animal that are large in other parts of the world are smaller here, while those that are usually small are often giants in the Neotropical zone. The llama, pudu, tapir, peccary, puma, jaguar, spectacled bear, rhea, and the monkeys are all small compared to their counterparts.

There is, however, one group of mammals in the Neotropical zone which cannot be compared with those in other zones, for they are peculiar to it: the Xenarthra, consisting of the anteaters, sloths, and armadillos. They vary considerably in size. The giant anteater (*Myrmecophaga jubata*) is sometimes nearly 8 feet long—thanks to its enormous plume of a tail which is as long as the rest of its body. The pygmy anteater (*Cyclopes didactylus*) is hardly as big as a squirrel. The armadillos range from the giant armadillo (*Priodontes gigas*), 5 feet long, to the *pichiciago* (*Chlamyphorus truncatus*), a mere 6 inches. The sloths, the most phlegmatic of mammals, are never larger than a young chimpanzee.

44. Giant anteater.

Nevertheless there are legends in South America about quite enormous beasts. In Patagonia travelers have collected from the Tehuelche Indians a description of a large nocturnal animal, of which no trace can be found among the living fauna of the continent. It is a large heavy beast, as big as an ox but with shorter legs and covered with thick, short, coarse hair. It is armed with enormous hooked claws, just like an anteater's, with which it digs a huge burrow, where it sleeps all day: that is why it is hardly ever seen. According to some versions its habits are amphibious. All agree that it cannot be harmed by arrows or bullets, and oddly enough, this last fabulous-looking characteristic is the best evidence that the monster is not a fable.

Moreover South America was not always inhabited only by relatively small mammals. In 1789 gigantic bones which seemed to belong to an animal almost as large as an elephant were found in mud on the banks of the Rio Luyan not far from Buenos Aires. The peons were the only people not to be surprised at this discovery: it was a giant mole, they said, which died whenever it had the misfortune to be exposed to sunlight. This naive explanation satisfied no one else, so the Viceroy of the colony had the bones sent to King Charles IV of Spain, where they were entrusted to José Garriga. This naturalist at once set about preparing, assembling, and mounting the skeleton so that he could make a detailed description. The results of his slow and painstaking work were published in 1796, to the great excitement of the entire scientific world. Goethe wrote an essay on the monster in which some have seen the beginnings of a theory of evolution.

Meanwhile a French diplomat had visited the Spanish scientist and had taken to the Paris Museum five remarkable engravings of this skeleton. A young naturalist called Georges Cuvier set about placing the monster in its zoological classification on the mere evidence of these drawings. In a communication to the Académie des Sciences he gave the name of *Megatherium* ("large mammal") to the animal which he recognized as a sort of "giant sloth."

> Its teeth prove that it lived on vegetables, and its sturdy fore-feet, armed with cutting claws, lead us to believe that it was principally their roots that it attacked. Its size and its foreclaws would have provided sufficient means of defense. It could not run fast, but it was not necessary for it to do so, as it had no need either to pursue or to flee . . .
>
> Its similarities relate it to several kinds of edentates. It has the head and shoulders of the sloths. The legs and feet provide a strange mixture of characteristics belonging to the anteaters and the armadillos.

We can now see that there is nothing monstrous in this mixture, since all the ingredients come from very closely related animals. But everyone was surprised at the size of this South American animal. When it stood on its hind legs it was about 15 feet high.

Unlike present-day sloths, it could not have lived in trees. From all that we now know about it, it seems to have been a sort of huge and hairy bear, armed with terrible claws. It must have usually stood upon its hind legs. Its heavy and powerful tail, the structure of its pelvis and hind legs forming an enormous supporting base, the type of articulation of its vertebrae, all show a clear tendency to an upright posture. Subsequently, at Carson City in Nevada, huge footprints, 18 inches long by 8 inches wide, were found to confirm this theory. If the creature sometimes put its forefeet on the ground (like the kangaroo when it walks without jumping), it must have been so lightly that the heavy hind feet at once obliterated all trace of them.

The Megatherium's great height enabled it to feed on the foliage of trees, hooking its claws firmly onto their branches so that it could browse lazily with the least effort.

Naturalists at the end of the eighteenth century and for most of the nineteenth did not dream that it could possibly have survived until our time. In Europe the large animals had been extinct since a geological era which according to Cuvier's Theory of Revolutions of the Globe had preceded our own. On this analogy the Megatherium was thought to have perished in the Tertiary era. Man could never have known this monster of a past age.

This did not prevent King Charles IV from ordering the officials of his transatlantic colony to send him one dead or alive. Several generations of scientists laughed at the king's folly. But since then much evidence has shown that the king was nearer to the truth than the scientists. Today no zoologist dares swear that it was impossible to grant at least part of the king's wish and send him the body not of an actual Megatherium, but of a relative almost as large. For a Megatherium is not the only giant in the group of sloths, which collectively are known as gravigrades. They were represented by a whole series of giants, among which were the Mylodons (*Mylodon, Glossotherium, Grypotherium*) that were very like

45. Reconstruction of the Megatherium.

the Megatherium, except that they were about the size of an ox and had much longer tails. Little by little as the years went by it became evident that these beasts, though huge and fantastic, were not such old fossils as the extinct pachyderms in Europe, which scientists eventually had to admit had been contemporary with man, when pictures of them were found in caves. Given the clearly recent date of the deposits in which the remains of the Megatherium and other no less impressive creatures were found, they could not have lived more than 30,000 years ago. But as it was almost universally believed that man did not invade America, by way of the Bering Strait, until quite recently—between 2,000 and 20,000 years ago according to various estimates (and in the nineteenth century scientists inclined to the shorter period *)—no one thought at first that

* It now seems certain that the oldest human remains in America are between 10,000 and 15,000 years old.

human beings could have known these monsters. And no prehistorian had ever been able to show the smallest drawing of the giant edentates comparable to the French cave paintings of mammoths and reindeer.

But in 1870 in a terrace of the little valley of the Rio Frias near Mercedes the famous Argentine paleontologist Florentino Ameghino discovered some human bones, and with them were some cut stones, charcoal, and some animal bones that had been pierced or incised, the remains of another huge monster: the Glyptodon (see Fig. 52, page 179). This was not a giant sloth but a giant armadillo with a rigid carapace like a tortoise's shell, which alone was sometimes as much as 12 feet long.

In 1881, near the Rio Arrecifer, Professor Santiago Roth found an almost complete human skeleton curled up by a Glyptodon's carapace. Further similar finds soon showed that the first South Americans used Glyptodon carapaces either as houses or as tombs.

I need hardly say that these discoveries were bitterly disputed. Anthropologists refused to believe that man could have come to America as *early* as the date when giant edentates were supposed to have lived. Paleontologists denied that these creatures could have survived so *late* as man's arrival in the continent. Both proved to be wrong. One day a giant sloth's skeleton was found at the bottom of what had once been a pit. The four legs were still complete, but all the middle of the body had been more or less destroyed and had been replaced by charcoal and ashes. Obviously the sloth had been trapped in the pit, and as they were unable to get it out the Indians had roasted it *in situ* by lighting a fire under its belly.

There was now no doubt that the Megatherium had lived at the same time as man. But for how long had it continued to do so? Man's first migrations in Alaska seem to have happened about 12,000 years ago, around 10,000 B.C. But when did the first men reach the very tip of South America? Several thousand years later, no doubt—let us say 7,000 B.C.* Thus the first American Indians could equally well have caught giant edentates and eaten their flesh in historical times or several thousand years earlier. The date the monsters had become extinct was still very much in doubt—if they really were extinct.

Then, in 1898, someone brought Ameghino a lot of little bones the size of beans from southern Patagonia and asked him what

* In the cave of Palli-Aike, near the Straits of Magellan, hearths have been found containing half-charred bones of horses, guanacos, and Mylodons. Thanks to the carbon 14 method Libby and Arnold have recently been able to fix their age exactly at 8,700 years.

animal they could have come from. They had been extracted from a piece of leather ¾ inch thick and covered with long reddish-gray hairs in which they had been embedded like small cobbles in a road. This skin belonged without the slightest doubt to a large ground sloth, and Ameghino was even able to determine its origin more exactly. It was already known from paleontological discoveries that some of the giant gravigrades were protected all over with these little bones. From the size and shape of the bony nodules, he reckoned that they had come from the fresh remains of an animal similar to a Mylodon.

Ameghino boldly published a first note upon "a living representative of the ancient fossil gravigrade edentates of the Argentine." He had a good reason for being so positive. A friend of his called Ramon Lista, who was the Argentine Secretary of State, had seen a

46. Asian pangolin.

curiously similar animal several years before in southern Patagonia. It resembled an Asian pangolin (*Manis*) both in shape and size, except that instead of being covered with scales it had long reddish-gray hairs. They shot at the beast several times, but this did not seem to trouble it in the least, and it vanished into the bush apparently unharmed.

Ameghino concluded that the gravigrade from which his bony nodules had come was of the same species as that seen by his friend. It must be a close relation to the Mylodon, a long-tailed ground sloth remarkably similar in outline to a pangolin. The Mylodon, while not so vast as the Megatherium, was by no means so small as the Indian pangolin, which is never more than 4 feet 3 inches in length, of which 1 foot 9 inches is tail alone. But as the Tehuelche spoke of a rather large animal, might not Lista have seen a young specimen? At all events the possible existence of a surviving ground sloth of small size was of immense zoological interest, and in his original 1898 description, Ameghino never identified the mysterious creature with the Mylodon, which was as big as an ox. Lista had just been killed by the Tobas Indians while he was exploring the Pilcomayo region,* and Ameghino named this new little Mylodon *Neomylodon listai* in his honor.

Everyone was eager for more news of the mysterious creature. The first thing to be found was the piece of fresh hide which he had referred to—though not very explicitly. This did not take long, for similar pieces of skin had come into the hands of several scientists: the Swedish explorer, Dr. Otto Nordenskjöld in 1896 and Dr. Francisco P. Moreno, Director of the La Plata Museum, in 1897.

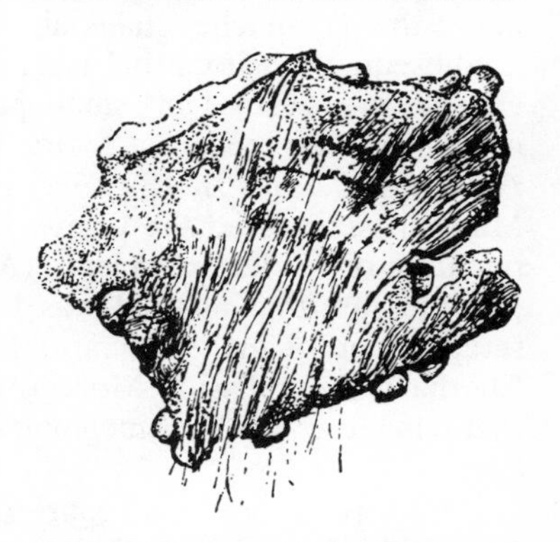

47. Piece of Mylodon's hide, stll bearing hair and small bones in the skin. (Paleontology gallery of the Paris Museum.)

These samples had both been cut from a piece of leather 5 feet long by about 2 feet 6 inches wide which had been casually hung from a tree to mark the boundary of an *estancia* belonging to a retired German captain called Eberhardt, who had found the skin in January 1895 while he was exploring a cave with several companions. This cave, as vast as a cathedral, in which they also found a human skeleton in a little niche, lay on the shores of a fiord called Ultima Esperanza—Last Hope Inlet. The skin, which was found buried in the ground, carefully rolled up inside out, had clearly been prepared by human hands.

Nordenskjöld visited Eberhardt in 1896 and systematically excavated the Cueva Eberhardt and brought to light the horny sheath of a ground sloth's enormous claw. He took his finds back to Sweden, where Einar Lönnberg published a description of them in 1899. This zoologist agreed with Ameghino about the animal the skin had come from: the bony nodules, and also the claw, showed that it was related to the Mylodon. But, from the size of the whole skin, the *Neomylodon* must be about the size of a small rhinoceros—which happened to be the size of one of the Mylodons whose remains were found in pampean deposits. He thought that

* And not by political enemies as Willy Ley reports, presumably under the impression that this is the normal death of all South American politicians.

the creature Lista had seen must have been a quite different animal. But this does not necessarily follow, since even a rhinoceros is no bigger than a calf at one stage of its existence. Finally he pronounced that the animal was certainly no longer extant:

> It is absolutely impossible to think that this animal, if it was still among living beings, could have eluded the sharp eyes of the native Indians.

This opinion is all the more surprising since Ameghino had already noted the Tehuelches' tales about the animal.

Subsequent visitors to Eberhardt's *estancia,* most of them Chilean officers, carried off small pieces of the skin, and presumably it was one of these souvenirs that eventually found its way to Ameghino's desk. In 1897 Dr. Francisco Moreno, of the La Plata Museum, excavated the Cueva Eberhardt, finding nothing but the remains of a large rodent (*Miolania*). At Eberhardt's farm, Moreno saw that all that was left of the giant sloth's skin was a scrap about 18 inches square. This was better than nothing. With Eberhardt's permission Moreno took it back to the Museum of La Plata in July 1898. Ameghino's bombshell burst a few months later.

Dr. Moreno took this square of skin to London, and on January 17, 1899, showed it to the members of the Zoological Society of London, who could hardly contain their excitement. Nevertheless he claimed not to be astonished by his examination of the skin, which was obviously fresh and had been cut off by human hands. For it had now been established that Mylodons were contemporary with man farther north in the Argentine. Moreover Moreno had himself found in a neighboring cave another mummified body, also very well preserved, of a man of an extinct race quite unknown to the Tehuelche who now inhabited the area. He therefore maintained, with Professor Seeley's support, that the Mylodon's skin could have remained in an apparently fresh state for untold ages.

The whole meeting disagreed with him and were sure that the skin belonged to a recently dead animal; it was still fairly supple, and one could see remains of muscles and ligaments and a "coating of dried serum was still preserved on the old cut edges," as Sir Arthur Smith Woodward who undertook the official description remarked. Sir Arthur "would, indeed, have unhesitatingly pronounced the skin to belong to a recent animal killed quite lately, had not Dr. Moreno been able to give so circumstantial an account of its discovery."

It was at this point that Professor Ameghino's second note appeared. In it he tells how his brother Carlos had sent him some little bones in 1897, saying that the Indians said that they came from an animal they called *iemisch* or in Spanish *tigre de agua* (water tiger). Carlos wrote:

> This animal is of nocturnal habits, and it is said to be so strong that it can seize horses with its claws and drag them to the bottom of the water. According to the description I have been given, it has a short head, big canine teeth, and no external ears; its feet are short and plantigrade, with three toes on the forefeet and four on the hind; * these toes are joined by a membrane for swimming, and are also armed with formidable claws. Its tail is long, flat, and prehensile. Its body is covered with short hair, coarse and stiff, of a uniform bay color. Its size is said to be larger than a puma's, but its paws are shorter and its body thicker.

Ameghino seems to have gone completely astray in accepting this theory of his brother's that the armored skin came from the Patagonians' *iemisch*, which hardly agrees with one's idea of a ground sloth like a Mylodon. A monster of this kind might have been rather frightening because of its size and its claws, but it is too much to believe that this undoubted vegetarian seized horses and drowned them, let alone that it should have canine teeth, or indeed that an edentate should have ordinary teeth that looked like canines. The *iemisch*'s alleged amphibious habits—which its webbed feet confirm—are a long way from the traditional picture of the Mylodon, and it would be very strange for a water animal to have huge burrowing claws.

Meanwhile several more scientific expeditions had been to the Cueva Eberhardt. The first was led by Erland Nordenskjöld, cousin of the Otto already mentioned, who excavated the cave and brought back a large quantity of the little bones mixed with dung and finely chopped hay which constituted the soil in the place where the skin had been found. After a very thorough comparative study Nordenskjöld remarked that the little bones and the skin itself could be attributed to a known Mylodon, *Glossotherium darwini*, whose fossil remains had been found in pampean deposits and described by Sir Richard Owen. Assuming, for lack of information to the contrary, that Ameghino's bones had come from the Cueva Eberhardt, his name of *Neomylodon listai* was a needless synonym for *Glossotherium darwini*. This was a blow to the professor, who had

* Professor Ameghino later altered this to four foretoes and three hind toes, to agree with the anatomy of the Megatherium.

a passion for naming animals, and had baptized such astronomical numbers, especially of fossils, sometimes hastily and often mistakenly, that many paleontologists refused to accept his names for them. After Nordenskjöld's statement he hastened to publish his second note with its not entirely convincing story that his bones had not come from the Cueva Eberhardt.

This suspect behavior does not mean that Lista was lying, or that the Eberhardt skin was not strangely fresh. Albert Gaudry, the famous French paleontologist, wrote:

> The pieces of skin which M. Lönnberg showed me at Uppsala with their hairs still firmly attached, a bone with dried muscles still clinging to it . . . droppings, finely chopped hay in a fresh state, and several horny parts of claws intact, are inexplicable unless the *Neomylodon* took refuge in the Cueva Eberhardt at a fairly recent date. There are no reasons for doubting M. Ameghino's belief that one might find it alive.

After the Swedes an Argentine expedition under the geologist Rudolf Hauthal, Dr. Moreno's assistant, set out in April 1899 and put the contents of the cave of Ultima Esperanza through a fine sieve. Eberhardt's skin had not been found in the entrance of the cave, but in an inner chamber, where the human skeleton had also been found. A crude wall of boulders closed the entrance except for a narrow passage. Some 50 yards inside the chamber a second and very thick wall barred the way and formed a raised platform. In the middle of this part of the chamber rose a little artificial mound. The ground was covered with a layer of dust and pebbles varying between 1 and 3 feet in depth, in which were found obvious remains of a kitchen: mussel shells and charred pieces of guanaco and deer bones. Under this superficial layer, near the platform, they came upon a great mass of excrement belonging to some herbivorous animal: part of it had been burned and reduced to ashes. Nearer the central mound they dug up a heap of dry hay in a good state of preservation. In this lower layer of dung, fodder, and rubbish they found numerous broken bones of the so-called *Neomylodon* (alias *Glossotherium*), as well as remains of a fossil horse and a large unknown carnivore. Finally, near the place where the first rolled-up skin was found, they found another 3 feet 8 inches by 3 feet in size.

The peculiar arrangement of the place, the skeleton that had already been found there, the kitchen midden, the various walls, the heaps of dung and fodder, 3 feet deep in some places, all seemed to show that giant sloths had been kept alive in this special enclosed part of the cave. The excrements were obviously those of a large animal, since the best-preserved ones had kept their shape and were almost as large as an elephant's. Moreover, an examination of the skull of a sloth revealed that man must have brained it with large stones. Clean-cut bones showed that the beasts had then been dismembered with sharp cutting tools and were, of course, finally eaten.

> The men who lived there ages ago [concludes Dr. Hauthal] were accustomed to stable their domestic animals in this part of the cavern, reserving the rest for their own dwelling place.

Subsequent analysis of the excrements by Spencer Moore showed that the animal's food was too cleanly cut at the end to have been done only by its blunt teeth. It must have eaten fodder that had been reaped by man.

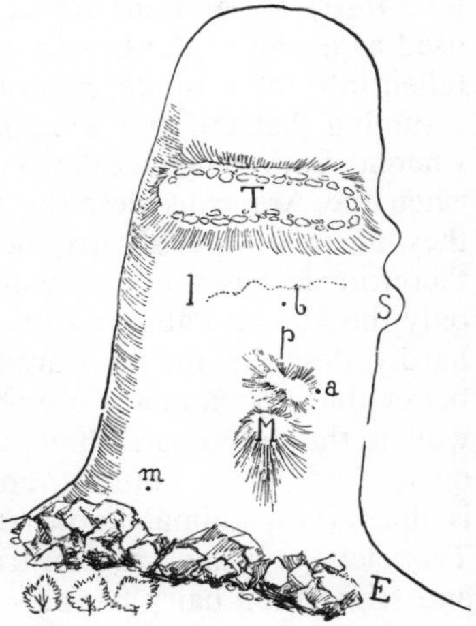

Can one call these sloths true domestic animals? It is far more likely that men used to drive them into such caves, just as in India wild elephants are caught by driving them into stockades, than that they were really domesticated. If the sloths were actually domesticated by the Indians—that is to say, they bred in captivity—why does this happen no longer? No other instance is known of an animal which has ceased to be domesticated, and no species of animal is less likely to become extinct than a domesticated one—and with good reason.

But is the ground sloth really extinct? What was the "hairy pangolin" that Ramon Lista vainly shot at? And we must not forget that several people reported Tehuelche legends not about a large, savage, and amphibious

48. Plan of the chamber in the Cueva Eberhardt, where the two rolled-up Mylodon skins (*a, b*) were found. *E:* narrow entrance left at the side of a rough boulder wall. *S:* niche containing human skeleton. *T:* raised platform. *M:* small mound. *p:* chopped hay. *l* (dotted line): line where ashes and layer of dung meet. *m:* mussel shells (after R. Hauthal).

carnivore, but a burrowing and harmless animal that was said to be invulnerable. Its description agrees well with one's idea of a sloth which had not, like the ai and unau, decided to live in trees.

Dr. Moreno reports that the Tehuelche and the Gennake tell of a sort of terrifying hairy beast, which is very rare, and an old chief showed him a cave that was said to be the lair of one of these animals. But the Indians never told him that the animal still lived.

Ramon Lista and Professor Santiago Roth both report the Tehuelche belief in a large and terrifying beast. Rudolf Lenz reports that in an Araucan epic there appears a wild beast whose name the Araucans translate into Spanish as *lofo-toro* (that is *lobo-toro*) or, as Lenz translates it, "wolf-bull," the "wolf" being used as a sign of the beast's savagery. But he seems to me to have fallen into the mistake, natural to a German or an Englishman, of assuming that the first word is an adjective qualifying the second, whereas Latin languages are constructed the other way round. Thus when the Araucans describe the otter in Spanish as *zorro-vibora* they mean a viperish fox, not a foxy viper. The name *lofo-toro* therefore means a bovine wolf, and not a wolfish bull. If one has only the known Patagonian animals as terms of reference, one can hardly describe the appearance of a Megatherium or Mylodon better than by calling it a wolf as big and heavy as a bull, for the wolf is the only coarse-haired and rather shaggy beast, fairly low on its feet, in the country, and the bull introduced by the whites is the largest animal known there. There is little doubt that the Tehuelche and the Araucan traditions include a memory of a large and frightening hairy beast.

Admittedly there is nothing very frightening about the domesticated or half-domesticated state in which gravigrades lived in the cave at Ultima Esperanza. But the people who kept Mylodons in these stables were not Tehuelche or Araucan but a preceding race, and in India many peoples are terrified of wild elephants, although others tame them. From what we know of the present and the subfossil fauna of this area the legendary beast can only be some kind of large ground sloth.

Professor Ameghino was not always so misguided as when he introduced his *iemisch* into the controversy. After the publication of his first note, while other scientists were excavating the Cueva Eberhardt, he began burrowing no less busily into the early accounts of the discovery of the New World. In the *Historia de la Conquista del Paraguay, Rio de la Plata y Tucumán* (1740-46) by a Portuguese Jesuit called Pedro Lozano he read of a large and very savage beast called *su* or *succarath*, which was in the habit of carrying its young on its back. The natives were said to hunt it and makes cloaks out of its skin. Therefore, thought Ameghino, it must be too thick and stiff to be used for finer garments; this all fitted the Eberhardt Mylodon. Its method of carrying its young is a characteristic of existing sloths and also of the anteaters which, apart from their size, fit the *succarath* fairly well.

In fact this beast had already been described by Gesner after the account given by André Thévet. Father Thévet, who was one of the most brilliant and learned minds of the French Renaissance, published in 1558 one of the first books in French about America, in which he describes many American animals with sober exactitude, of the region "of the Straits of Magellan," he says:

> This region is of the same temperature as Canada and other countries that approach our Pole: thus the inhabitants dress themselves in the skins of certain beasts, which they call in their language *su,* which is as much as to say water: for, in my judgment, this animal lives for most of the time on the banks of rivers. This beast is very wonderful, made in a very strange fashion, wherefore I have chosen to show it in a picture. Another thing: if it is pursued, as many of the people of this country do in order to have its skin, it takes its young on its back, and covering them with its long and large tail, escapes by fleeing with them. However the Savages use a trick to catch this beast: making a deep pit near the place where it is in the habit of making its abode, and covering it with green leaves, so that when running, without suspecting the ambush, the poor beast falls in this pit with its young. When it sees that it is caught, it maims and kills its young (as if maddened): and gives such terrible cries that it makes the Savages very fearful and timid. Yet in the end they kill it with arrows and then they flay it.

The naive picture which accompanies the text shows a very emaciated sort of lion with a plume of a tail like an anteater's and a grotesque head somewhat reminiscent of a bearded man. On its back crouch half a dozen young ones. It is more than likely that this picture was not drawn by Thévet himself but by someone else relying on his verbal description, for this was the usual practice at this time, and explains why so many animals which are correctly described in the text are so strangely distorted in the pictures. The text, for instance, makes it quite clear that the bison is a sort of hairy and humpbacked bull, but the picture is not in the least like one. If we bear this in mind, our picture could very well refer to one of the giant sloths, especially the Mylodon, which, unlike the Megatherium, had a long tail. It was no doubt because the artist was told it had claws that he gave it paws like a lion. The human-looking head shows that it was not an anteater—for they all have

49. Naive drawing of the *su* or *succarath,* after Father André Thévet (1558).

tube-shaped skulls. On the other hand, it could have been a sort of sloth. Compare it with a recent portrait of an ai drawn from nature by I. Cooper and you will see that the gravigrades have faces like a caricature of a man.

It seems quite likely that Father Thévet's *su* is some kind of ground sloth, and that these animals lived in Patagonia until a fairly recent date—let us say the Middle Ages.

At the turn of the century many scientists' skepticism had been much shaken by the accumulation of converging evidence: Ramon Lista's hairy pangolin—so very like a Mylodon; the discovery of such well-preserved remains; the Tehuelche and Araucan legends, and finally Father Lozano's story, whose original source I have just quoted. Even Sir Ray Lankester eventually admitted that large ground sloths might survive in some little-known or almost unexplored part of Patagonia. The *Daily Express* sponsored an expedition to bring back a living Mylodon, thus resuscitating King

Charles IV's command, this time with more encouragement from the scientists. But when the first searches yielded nothing, the leader of the expedition, H. V. Hesketh-Prichard, grew impatient before he had even reached Last Hope Inlet and went home to England in a huff, hardly concealing his disappointment and furious at his own failure, telling everyone that he had been hoaxed and that the Patagonian legends were a complete invention.

50. Portrait of an ai (after a recent water color by I. Cooper).

Today, half a century later, the matter is still at a dead end. Cavendish's and Koslowsky's expeditions came back as empty-handed as Hesketh-Prichard's, and scientists have therefore concluded that living Mylodons no longer exist. It has also been pointed out, very justly, that the Tehuelche accounts never come from direct witnesses but always refer to third parties, who doubtless never saw the fabulous beast either.

But is Patagonia the right place to look for any survivors of these animals? There is not the slightest doubt today that in South America man knew all these giant mammals that seem to belong to a vanished age. It is also well established, on irrefutable paleontological proof, that many of them once migrated to North America, where they were contemporary with man about 10,000 years ago. From the Miocene to the end of the Pleistocene, and even at the beginning of the present period, giant sloths spread north through Uruguay, Brazil, Bolivia, Colombia, Central America, Mexico, and even part of the United States, where their remains have been found in kitchen middens.

It is now proved beyond a doubt that the remains of Mylodons unearthed at the southern tip of Patagonia have been preserved in their cave for many centuries and that these huge and harmless beasts were exterminated there by the Indians who preceded the Tehuelche. The carbon 14 method of dating has fixed the age of the dung in the Cueva Eberhardt in which the Mylodon's bones and skin were found at about 10,000 years (to within 400 years either way). On the other hand, one may suspect that giant sloths —or at least smaller related ground species—were known farther north in the Argentine at the time of the conquest of America, as appears from the old chronicles. And perhaps some few survivors were responsible for the legend current among the Tehuelche during the last centuries and for Ramon Lista's more recent report.

No doubt the sloth empire covered its greatest area some 10,000 years ago. At this time there must have been a large number of intermediate forms between the giant species in the savanna and the little species in the forests. Slaughtered by the nomadic hunting Indians, both in the pampas of the south and the green prairies of the north, the largest sloths would have retreated, as the jaguar did, to the tropical forests, where they could find a safer refuge. All the same, it is unlikely that the really gigantic species could have adapted themselves to the inextricable virgin forests, the habitat in which the small tree species flourished. On the other hand, it is not difficult to see how the medium-sized ground sloths might have survived in wooded savanna or sparse forest, or even on the fringes of or in clearings in the densest of jungles. For the great ground sloths were not destroyed by any revolutionary geological or climatic change. From the number of their remains in kitchen middens it is clear that these large and peaceable beasts, like so many other species, were victims of man's gluttony. If such is the case, what has happened to them in their impenetrable retreat in the vast Amazonian *selva* and the *boscosa* of the Andes, through which they passed in the course of ages? It is hard to see what, in the peace of these forests rarely inhabited by man, could have led to their extinction. Only human traps were able to put an end to these armored brutes against which beasts of prey were powerless. Might they not still live in this "green hell" and find it a haven of peace? At all events, as we shall see, this would explain some of the rumors that are still current in an area where "nothin' would surprise."

CHAPTER **11**

THE GIANT ANACONDA AND OTHER INLAND "SEA SERPENTS"

Th'old Dragon under ground
In straiter limits bound . . .
. . . Swindges the scaly Horrour of his foulded tail.
JOHN MILTON, *On the Morning of Christ's Nativity*.

IN 1906 the Royal Geographical Society sent Major Percy Fawcett, an artillery officer of 39, to make a thorough survey of the area of the Rio Abuna and Acre River, a claim to which was disputed by Brazil, Bolivia, and Peru. It was to perform this thankless task that he first penetrated the vast forests of the Amazon, in which he was to vanish without trace 20 years later.

It was in January 1907 that he first heard of gigantic snakes.

> The manager at Yorongas [Fawcett writes in his memoirs] told me he killed an anaconda fifty-eight feet long in the Lower Amazon. I was inclined to look on this as an exaggeration at the time, but later, as I shall tell, we shot one even larger than that.

This happened two or three months later on the Rio Abuna, upstream of its junction with the Rio Rapirrão:

> We were drifting easily along in the sluggish current not far below the confluence of the Rio Negro when almost under the bow of the *igarité* there appeared a triangular head and several feet of undulating body. It was a giant anaconda. I sprang for my rifle as the creature began to make its way up the bank, and hardly waiting to aim smashed a .44 soft-nosed bullet into its spine, ten feet below the wicked head. At once there was a flurry of foam, and several heavy thumps against the boat's keel, shaking us as though we had run on a snag.
>
> With great difficulty I persuaded the Indian crew to turn in shorewards. They were so frightened that the whites showed all

51. Major Percy Fawcett's encounter with a 62-foot anaconda (from a drawing by his son Brian).

round their popping eyes, and in the moment of firing I had heard their terrified voices begging me not to shoot lest the monster destroy the boat and kill everyone on board, for not only do these creatures attack boats when injured, but also there is great danger from their mates.

We stepped ashore and approached the reptile with caution. It was out of action, but shivers ran up and down the body like puffs of wind on a mountain tarn. As far as it was possible to measure, a length of forty-five feet lay out of the water, and seventeen feet in it, making a total length of sixty-two feet. Its body was not thick for such a colossal length—not more than twelve inches in diameter—but it had probably been long without food. I tried to cut a piece out of the skin, but the beast was by no means dead and the sudden upheavals rather scared us. A penetrating foetid odour emanated from the snake, probably its breath, which is believed to have a stupefying effect, first attracting and later paralysing its prey.* Everything about this snake was repulsive.

Such large specimens as this may not be common, but the trails in the swamps reach a width of six feet and support the statements of Indians and rubber pickers that the anaconda sometimes reaches

* While snakes' breath has, of course, none of the stupefying properties with which it is often credited, it can clearly be very foul, like that of other carnivores. A lion tamer has wittily said that the chief risk in putting one's head in a lion's mouth is of being asphyxiated. Here it is more likely that the insufferable stench comes from the snake's cloacal musk glands which play some part in sexual attraction.

an incredible size, altogether dwarfing that shot by me. The Brazilian Boundary Commission told me of one killed in the Rio Paraguay exceeding eighty feet in length!

Fawcett was certainly a dreamer, and his dreams sometimes led him to cherish the wildest hopes, but he was not a liar. What he saw he always reported in a very matter-of-fact way; his interpretations are sometimes fantastic, but his observations never are. The most striking thing in his memoirs is the contrast between the sordid hell he describes and the splendor of the lost cities that he searched for till his death.

What then is the official record size for an anaconda (*Eunectes murinus*) in the zoological textbooks? Most of them will allow 30 feet. The American herpetologist Thomas Barbour, the great Brazilian expert Dr. Afranio do Amaral of the Institute at Butantan, and Dr. José Candido de Melo of the Rio de Janeiro Zoo all agree on 45 feet; but other experts are more cautious. Hyatt Verrill, for instance, writes:

> Personally I do not believe the anaconda ever attains a length of more than twenty feet. Dr. Ditmars is even more conservative and sets the limit at nineteen feet. But as I have personally killed anacondas which were within a fraction of an inch of twenty feet from tip of nose to tip of tail, I add one foot to this famous scientist's limit. But until I have seen a living anaconda, or an unstretched skin which actually measures more than twenty feet, and no one has produced such yet, I shall maintain twenty feet as the greatest length reached by this largest of American snakes.

A tanned snakeskin is always noticeably longer than it was on the living animal, since it has to be stretched in the drying. If it is unscrupulously overstretched it can be considerably lengthened. Hyatt Verrill claims to have seen the skin of an anaconda which measured only 18 feet alive but had been stretched to 23 feet. But the Institute at Butantan in Brazil has an anaconda skin 33 feet long. I do not think it could have been stretched to this length from only 20 feet, or that such a reputable scientific institution would have allowed it to be so overstretched. It must have come from a snake at least 26 feet long. Besides, if we are to believe Curran and Kauffeld, the greatest American experts on snakes, the largest specimen which has been exhibited in a zoo measured 26 feet. And you can't stretch a living snake.

I should also add that at El Fayum in Egypt, parts of the spinal column of a python called *Gigantophis* belonging to the Middle

Eocene period were dug up; the entire animal was reckoned to be between 42 and 65 feet long. Fawcett's snake would not necessarily have been the longest.

Verrill maintains that the much exaggerated lengths that travelers and Indians often attribute to anacondas are due to the difficulty of judging the size of a living snake. One day he saw an anaconda curled up on a ledge of rock in Guiana and made the experiment of asking each of his companions to guess its length:

> The camera man, who had never before been in the jungle, said sixty feet. The missionary who had spent seven years in the interior and had seen scores of big snakes was more conservative and said thirty feet. The Indians' estimates varied from twenty to forty feet; my camp boy, who had accompanied a party of snake collectors a few years earlier, said thirty feet . . .
>
> A twenty-two rifle bullet through the head brought an end to the big anaconda's career and when he was straightened out and measured he proved to be exactly nineteen feet and six inches in length. But what a monster! About his middle he measured thirty-three inches and a fraction and he weighed over 360 pounds and was a heavy load for five Indians. So huge were his proportions that even after we had measured the creature it was difficult to believe that he was so "small."

Verrill's argument is sound enough, but is his experiment as instructive as he would have us think? The Indians who judged that a *coiled-up* snake was 20 feet long were remarkably accurate in their observation, while those who put it at 40 feet would surely have been much more accurate if they had seen the snake uncoiled. And even if, on the evidence of the missionary, we must halve the dimensions of snakes reported by people who know the country well, we shall still find many instances of specimens much larger than the record skin in the Institute at Butantan.

Fawcett could not have misjudged the length of his snake to any great extent, since he measured it dead on the riverbank. His phrase "as far as it was possible to measure" may mean that he could not measure the 17 feet in the water, but more probably that he had no exact means of measurement at hand and so in the usual British fashion paced out its length with his feet—and could not have gone so far wrong. If he had merely guessed its length he would not have given exact figures but have written "about 50 feet out of the water and 15 feet in the water."

Other more recent explorers of the Amazon country mention anacondas much longer than the 20 feet that Verrill and Ditmars grudgingly concede. Let us begin with the Marquis de Wavrin. I was to have gone with him in 1940 as zoologist on his attempt to find the sources of the Orinoco, a project which was foiled by the war. I can confirm that he can be trusted to speak the truth. He says that the large type of anacondas "generally met with along the rivers measure between 20 and 25 feet long. I have seen some over 30 feet long and there are even larger ones, if the natives are to be believed."

He once shot an anaconda about 25 feet long; it fell into the water from the branch around which it was coiled, and he said he would like to fish up the body, which could still be seen quivering in some 3 feet of water. His canoemen replied that it was a waste of powder to shoot such a small snake and a waste of time to stop and pick it up. They also told him:

> On the Rio Guaviare, during floods, chiefly in certain lagoons in the neighborhood, and even near the confluence of this stream, we often see snakes which are more than double the size of the one you have just shot. They are often thicker than our canoe.

In seven years on the Amazon, Up de Graff encountered only one of these gigantic anacondas. It lay in shallow water beneath his canoe:

> It measured fifty feet for certainty, and probably nearer sixty. This I know from the position in which it lay. Our canoe was a twenty-four footer; the snake's head was ten or twelve feet beyond the bow; its tail was a good four feet beyond the stern; the center of its body was looped up into a huge S, whose length was the length of our dugout and whose breadth was a good five feet.

Algot Lange, another explorer of the Amazon, even claims he shot an anaconda of about the same size: it was 56 feet long and had a diameter of 2 feet 1 inch. Willard Price says that Lange took its skin to New York, but surely the zoologists would have heard of this?

The stories of anacondas 70 feet long seem to be quite incredible. Yet I have heard from the witness's own mouth a circumstantial account of how a specimen of this astonishing size was killed. Having questioned and cross-questioned my informant for several days, I am as convinced of his sincerity as if I had witnessed the incident myself.

In 1947, after the Chavantes had massacred several Brazilian officials, Francisco Meirelles, of the Service for the Protection of the Indians, organized a new expedition to try to establish peaceful relations with this wild tribe on the Middle Araguaya. The party of some 20 men included two Frenchmen: the very young explorer Raymond Maufrais, who disappeared in Guiana, and the painter Serge Bonacase of Paris, who told me this story.

A party of seven or eight men had gone hunting capybaras in the swampy area between the Rio Manso and the Rio Cristalino.

> The guide [Serge Bonacase told me] pointed out an anaconda asleep on a rise in the ground and half hidden among the grass. We approached to within 20 yards of it and fired our rifles at it several times. It tried to make off, all in convulsions, but we caught up with it after 20 or 30 yards and finished it off. Only then did we realize how enormous it was; when we walked along the whole length of its body it seemed as if it would never end. What struck me most was its enormous head, which was like this. [He stretched out his arms in front of him with his hands together, thus forming a triangle with 2-foot sides and an 18-inch base.]
>
> As we had no measuring instruments, one of us took a piece of string and held it between the ends of the fingers of one hand and the other shoulder to mark off a length of 1 meter. Actually it could have been a little less. We measured the snake several times with this piece of string and always made it 24 or 25 times as long as the string. The reptile must therefore have been nearly 23 meters long [75 feet].

Even if the piece of string was only 90 centimeters (3 feet) long —and I don't think we can allow a margin of error of more than 10 per cent—the snake would have been between 72 and 75 feet long.

Serge Bonacase added that the animal's diameter was almost half the length of the measuring cord, probably nearly 18 inches. The beast was so heavy that they could not lift it up in the middle of its body. They could only guess at its weight.

No doubt the reader will wonder, as I did, why they did not take photographs of such a trophy. The reason is simple. The Service for the Protection of the Indians had forbidden them to take cameras—which could easily have terrified the Chavantes and caused another massacre—on their particularly tricky mission.

But why didn't they think of bringing back its skin or its head? Serge Bonacase's answer to this question was:

> First of all, none of us seemed to realize that there was anything exceptional about our prize. There were no zoologists among us. The

Brazilian officials who had spent much of their lives in this country did not seem to be particularly surprised. As for me, I had heard so many tales of giant snakes that I supposed the whole of the Amazon was crawling with monsters of this size.

Of course, we should have liked to take the snake's skin back, but we had neither the time to skin the beast nor to prepare its hide. We were also very tired, and in that country you must never let yourself linger by the way or encumber yourself with extra luggage. Think what a piece of skin more than 60 feet long by 4 feet 6 inches wide would have weighed! We should have had our work cut out to carry it—or for that matter the head. Besides, who would have been crazy enough to lug a piece of rotting meat on his back in that heat through country infested with insects?

It was undoubtedly an anaconda of the known species. "It was a very deep dark brown," Bonacase told me, "marked on the sides with almost black irregular oval rings."

Let us proceed to what may be another kind of snake. Lorenz Hagenbeck, late director of the Hamburg Zoo, was convinced of the existence of a kind of South American water snake larger than the largest anaconda and of quite staggering size. The son of the great Carl, he was brought up in the business. For more than a century travelers and animal catchers sent zoological information from all over the world to the Hagenbeck dynasty, who were proud to supply zoos in every country with specimens of rare and sometimes even supposedly extinct animals. It was an explorer sent out by a Hagenbeck who caught the first pygmy hippopotamus. But the Hagenbeck family papers also include several reports, transcripts of which I have seen, about an aquatic snake from the Amazon which is much larger than the largest anaconda and which the Indians seem also to distinguish by the name of *sucuriju gigante* or "giant boa."

Lorenz Hagenbeck was a close friend of the two priests, Father Heinz and Father Frickel, who reported it, and never doubted their sincerity for a moment. The first report comes from Father Victor Heinz himself:

> During the great floods of 1922, on May 22 at about 3 o'clock to be exact, I was being taken home by canoe on the Amazon from Obidos; suddenly I noticed something surprising in midstream. I distinctly recognised a giant water-snake at a distance of some 30 yards. . . .
>
> Coiled up in two rings the monster drifted quietly and gently down-

stream. . . . I reckoned that its body was as thick as an oil-drum and that its visible length was some 80 feet. When we were far enough away and my boatmen dared to speak again they said that the monster would have crushed us like a box of matches if it had not previously consumed several large capybaras.

A day's march south of Obidos one of these monsters was killed, it seems, just as it was swallowing a capybara on the muddy shore of Lago Grande do Salea. Its stomach contained no less than four adult specimens of these huge rodents. Elsewhere two large round excrements were found full of animal hair and with a bone from an ox's foot sticking out of one of them. Father Heinz also writes:

My second encounter with a giant water-snake took place on 29 October 1929. To escape the great heat I had decided to go down the river at about 7 p.m. in the direction of Alemquer. At about midnight, we found ourselves above the mouth of the Piaba when my crew, seized with a sudden fear, began to row hard towards the shore.

"What is it?" I cried, sitting up.

"There, a big animal," they muttered, very excited.

At the same moment I heard the water move as if a steamboat had passed. I immediately noticed several feet above the surface of the water two bluish-green lights like the navigation lights on the bridge of a river boat, and shouted:

"No, look, it's the steamer! Row to the side so that it doesn't upset us."

"Que vapor que nada," they replied. *"Una cobra grande!"*

Petrified, we all watched the monster approach; it avoided us and recrossed the river in less than a minute, a crossing which would have taken us in calm water ten to fifteen times as long. On the safety of dry land we took courage and shouted to attract attention to the snake. At this very moment a human figure began to wave an oil-lamp on the other shore, thinking no doubt, that someone was in danger. Almost at once the snake rose on the surface and we were able to appreciate clearly the difference between the light of the lamp and the phosphorescent light of the monster's eyes. Later, on my return, the inhabitants of this place assured me that above the mouth of the Piaba there dwelt a *sucuriju gigante.*

Father Heinz now began to study the subject with fervor, and to collect the reports which he eventually sent to Hagenbeck. One witness was a Portuguese merchant called Reymondo Zima, well known to Father Heinz.

On 6 July 1930 I was going up the Jamunda in company with my wife and the boy who looks after my motor-boat. Night was falling when we saw a light on the right bank. In the belief that it was the house I was looking for I steered towards the light and switched on my searchlight. But then suddenly we noticed that the light was charging towards us at an incredible speed. A huge wave lifted the bow of the boat and almost made it capsize. My wife screamed in terror. At the same moment we made out the shape of a giant snake rising out of the water and performing a St. Vitus's dance around the boat. After which the monster crossed this tributary of the Amazon about half a mile wide at fabulous speed, leaving a huge wake, larger than any of the steamboats make at full speed. The waves hit our 43-foot boat with such force that at every moment we were in danger of capsizing. I opened my motor flat out and made for dry land. Owing to the understandable excitement at the time it was not possible for me to reckon the monster's length. I presume that as a result of a wound the animal lost one eye, since I saw only one light. I think the giant snake must have mistaken our search-light for the eye of one of his fellow-snakes.

At this same place during a motorboat journey in 1948 Paul Tarvalho, an old pupil of Father Heinz, saw a giant snake emerge from the water some 250 to 300 yards away; he reckoned it was about 150 feet long. For a moment it followed the boat, which made off at top speed.

The Franciscan Father Protesius Frickel noticed the head of a giant snake lying in the water near the bank of the Rio Trombetas. He disembarked above the spot and cautiously approached to within half a dozen paces of the beast. "Its eyes," he wrote, "were as large as plates."

Father Heinz gives us an idea of the titanic strength of such monsters:

On 27 September 1930, on an arm of water that leads from Lake Maruricana to the Rio Iguarapé, a Brazilian named João Penha was engaged in clearing the bank to make it easier for the turtles to come up and lay their eggs. At a certain moment, behind one of those floating barriers made of plants, tree-trunks and tangled branches, against which steamers of 500 tons often have to battle to force a passage, he saw two green lights.

Penha thought at first that it was some fisherman who was looking for eggs. But suddenly the whole barrier shook for 100 yards. He had to retreat hurriedly for a foaming wave 6 feet high struck the bank. Then he called his two sons, and all three of them saw a snake rising out of the water pushing the barrier in front of it for a distance of some 300 yards until the narrow arm of water was finally freed of it.

During all this time they could observe at leisure its phosphorescent eyes and the huge teeth of its lower jaw.

There are also two striking photographs, taken 15 years apart, and published in a Rio de Janeiro newspaper. Photographs can be faked, so to make quite certain Father Heinz sought out the proprietor of the Bazar Sportivo at Manaos who had developed them both. It may seem odd that both rolls of film should end up in the same shop so many years apart, but Manaos is the first really civilized place that anyone coming in from the Amazon basin would reach and it cannot have many shops equipped to develop photographs. At all events the proprietor assured the priest that the negatives had not been retouched.

The officials of the Brazil–Colombia Boundary Commission, who brought the first photograph to be developed in 1933, told him how they had killed a 98-foot snake 2 feet in diameter with a machine gun on the banks of the Rio Negro. The wounded animal had, it seems, risen up some 30 feet, smashing bushes and even small trees under its weight of two tons. Four men had been unable to lift its head.

The second photograph was taken in 1948. The snake, which was said to measure 115 feet in length, crawled ashore and hid in the old fortifications of Fort Tabatinga on the River Oiapoc in the Guaporé territory. It needed 500 machine-gun bullets to put paid to it. The speed with which bodies decompose in the tropics and the fact that its skin was of no commercial value may explain why it was pushed back in the stream at once.

On the strength of all these reports Hagenbeck asserted that the *sucuriju gigante* was not a myth, and that it might reach a length of 130 feet and a diameter of 2 feet 6 inches; its weight would be 5 tons. Its general color was dark chestnut, and its belly was spotted with a dirty white. Its eyes were surprisingly large, and their terrifying appearance was increased by their phosphorescence.

Apart from the usual skeptics who refuse to believe any of these stories, there are others who have tried to prove "scientifically" that such monsters are impossible.

Louis Marcellin has attacked the legends on mathematical grounds:

Let us be generous and allow a certain margin to these proportions:

> 40 to 45 feet as serious authors have supposed without however adducing the beginnings of a proof. We are still a long way from the 130 feet in question and even more from the actual weight of a snake 30 feet long which weighs 250 lbs. Let us allow an average of 9 lbs. per foot. A "monster" of 130 feet would weigh at the maximum half a ton.

M. Marcellin (who incidentally is not so generous as he pretends, since Verrill's 19½-foot snake weighed 360 pounds, an average of nearly 19 pounds per foot) is a poor mathematician. And Dr. José Candido de Melo has made a similar mistake in asserting that a 5-ton snake would have to be 500 feet long. The weight of a snake would be directly proportional to its length only if its thickness were constant, and no one has been so foolhardy as to pretend that the diameter of a 130-foot snake is the same as that of a 20-foot one. If the proportions of the two are the same, the weight depends upon the volume; that is, upon the *cube* of the length. Worked out correctly, the weight for M. Marcellin's snake would therefore be just under 7½ tons, and if it had the proportions of Verrill's it would be no less than 45 tons. Both these calculations assume that the proportions of these giant boas are the same as those of smaller snakes. Since Hagenbeck has given us the snake's *diameter* as well as its length we can go to work another way (calculating for ease of arithmetic in the metric system). His snake is 40 meters (130 feet) long, 80 centimeters (2 feet 6 inches) thick, and it weighs 5 metric tons (a little over 5 American tons). Now, since it is an aquatic snake, it must, like all other aquatic animals, have a specific gravity almost equal to that of water. In other words, every cubic meter of snake weighs the same as a cubic meter of water: 1 metric ton. Now assuming that it tapers considerably at the tail, and Hagenbeck's 80 cm. is a maximum diameter, let us consider the snake as a cylinder of, say, 50 cm. diameter; i.e., 25 cm. radius. Its weight is then $40 \times 0.25^2 \times \pi$, which is about 7¾ tons. By the same method we can say that if a perfectly cylindrical 40-meter snake were to weigh 5 tons, its diameter would be 39 cm.

All in all, I think 5 tons a *very reasonable* weight for a 130-foot snake.

If this is the normal weight for these monsters, it is less surprising that none of the people who claimed to have shot one should have brought back a specimen, or waited until the flesh had rotted off the bones to bring back a bone. In the tropics flesh putrefies quickly, and the stench that would rise from 5 tons of rotting flesh defies the imagination.

Five-ton snakes are not the only monsters reported from this part of the world. The Marquis de Wavrin writes that "Around the Upper Paraguay they give the name *miñocão* to a more or less fabulous snake."

The first scientist to be interested in the *miñocão* or, in its more usual Portuguese spelling, *minhocão,* was Fritz Müller (1821–97), a German embryologist who was living in Brazil. At first he took no notice of what seemed absurd stories about an animal 50 yards long and 15 feet across, covered in bony armor, overturning trees like blades of grass, shifting the courses of rivers, and turning dry land into fathomless marshes. But gradually he collected evidence from people who claimed to have seen it and spoke of it not as a monster but a very large unknown animal, until he wrote a report on the probable existence in Brazil of a huge amphibious creature. The word *minhocão* was, according to Müller, the superlative of *minhoca* (earthworm), and thus meant "giant earthworm." In the late 1860s it made a very remarkable appearance in the neighborhood of Lage. Some six miles from the town, Francisco de Amaral Varella saw a strange animal of gigantic size. It was 3 feet thick, but not very long, with a pig's snout, but he couldn't say whether it had any feet. He dared not attack it alone, but when he called his neighbors it lumbered off clumsily, leaving a trail of deep furrows about 3 feet wide before it disappeared into the ground.

Some weeks later, a similar trench which could have been made by the same animal was seen on the other side of Lage, nearly four miles from the first. They followed the animal's track which finally led under the roots of a large pine and then were lost in swampy ground.

Emil Odebrecht was surveying a route for a road from Itajahy into the uplands of the province of Santa Catarina, and in a broad, swampy plain his progress was held up by winding trenches along the course of a stream. They were too wide for him to step across, but not too wide for him to jump them.

In January in the early 1860s Antonio José Branco who lived on a tributary of the Rio dos Cachorros 6 miles from Curitibanos, and had been away with his whole family for eight days, came home to find the road undermined, huge heaps of earth thrown up, and a grooved track 10 feet wide and about half a mile long, ending in a swamp. It had turned a stream from its old course. The animal's route was mostly underground, and where it went down under the stream, several pine trees were overthrown. One of the trees with its bark rubbed off was still to be seen in 1877. Hundreds of people came from Curitibanos and other towns to see the *minhocão's* work.

One evening in 1849, after a long period of rains, a sound of falling rain was heard in João de Deos's house near the Rio dos Papagaios, in Parana province, but the sky was clear and starry. Next morning a large piece of land on the other side of a hillock was completely undermined: deep furrows led to a stony plateau where heaps of reddish-white clay showed the route the beast had taken into the bed of a stream that ran into the Papagaios. Three years later Lebino José dos Santos sought out the place, found the tracks still there, and concluded that there had been two animals some 6 to 10 feet thick.

Senhor Lebino also related that in the same district a Negro woman, who was going one morning to draw water, found the pool destroyed and saw an animal "as big as a house" crawling away on the ground. She called her neighbors, who arrived too late to see the monster itself, but they saw its track where it had passed over a rock and disappeared into deep water. In the same neighborhood a young man saw a big pine tree suddenly fall without visible cause. He hurried up, found the surrounding earth moving, and saw a huge wormlike black animal, "no longer than a lasso" (about 80 feet) with two movable horns on its head lying close to its body. It was wallowing in the mud.

> The accounts as to the size and appearance of the creature [Müller remarks] are very uncertain. It might be suspected to be a gigantic fish allied to *Lepidosiren* and *Ceratodus;* the "swine's snout," would show some resemblance to *Ceratodus,* while the horns on the body rather point to the front limbs of *Lepidosiren,* if these particulars can be depended upon. In any case it would be worth while to make further investigations about the Minhocão, and, if possible, to capture it for a zoological garden!

Fritz Müller is not the only authority for the *minhocão*, nor is his theory about its identity the only or the most likely one.

The *Gaceta de Nicaragua* of March 10, 1868, published a letter from Paulino Montenegro telling how during a journey to Concordia in February he heard that a gigantic snake had settled in a place called La Cuchilla. He went there with some friends and saw tracks which convinced him of the existence of some large animal. Five years earlier a sort of platform of earth had appeared for no apparent cause at the foot of a hill. A peasant trustingly planted some fruit trees on it, but in 1863 the ground collapsed, laying bare a large rock; yet there was no water to cause the subsidence. Little by little, trees were loosened, great oaks were even overturned, and huge chunks of rock thrown up until in December they barred the road between Chichiguas and San Rafael del Norte. The ground still showed many crevasses and in places it had caved in. It had clearly been undermined. When Montenegro arrived the last track was three days old, and revealed that there had been two animals. The track lay in loose ground, and one could see where the beast had pushed over an oak and then retreated apparently frightened at the crash. Two tracks led from there, one directly into a pool, the other larger one first went over some stony ground, where it was 4 feet deep, and then straight down into the same pool. Roots of trees on the way were worn through and rocks weighing more than a ton and a half had been pushed aside. The animals seemed to have scales from the imprints they had left in the mud. Montenegro thought they must have been about 40 feet long (but how can one judge *length* from a groove in the ground?), 10 feet high, and 5 feet wide. They had been known locally as *sierpe*, serpents.

Fawcett also seems to refer to the *minhocão* when he writes:

> They talk here of another river monster—fish or beaver—which can in a single night tear out a huge section of river bank. The Indians report the tracks of some gigantic animal in the swamps bordering the river, but allege that it has never been seen.

The reports are usually rather vague about the animal's size and shape. Indeed it has been assumed on too little evidence to be a monstrous earthworm, merely because it leaves grooved tracks like a worm's. It might well have feet concealed below the earth or mud, although no observer mentions them. Dr. Budde, who summarizes Müller's report in his *Naturwissenschaftliche Plaudereien,* thinks the *minhocão* could well be some kind of gigantic armadillo, one of those alarming living tanks that I have mentioned in the previous chapter. The bony carapace sometimes mentioned, the snub nose and the horns—which are merely the upstanding ears—all recall some kind of Glyptodon. Given that this animal, which had powerful claws, was a burrower like the present armadillos, to which it is related, the picture is in some respects a very good likeness. As it was over 11 feet long the passages it dug must have caused considerable subsidence in the ground. In Africa the burrows dug by the aardvark, a much smaller animal, often make whole slices of road collapse and hold up the traffic.

Only one of the *minhocão*'s habits seems surprising in a Glyptodon: it is amphibious. Armadillos certainly are not, though it is now known that they are excellent swimmers. They cross narrow streams by running across the bottom, and for larger rivers they fill their intestines with air until they float in spite of their heavy

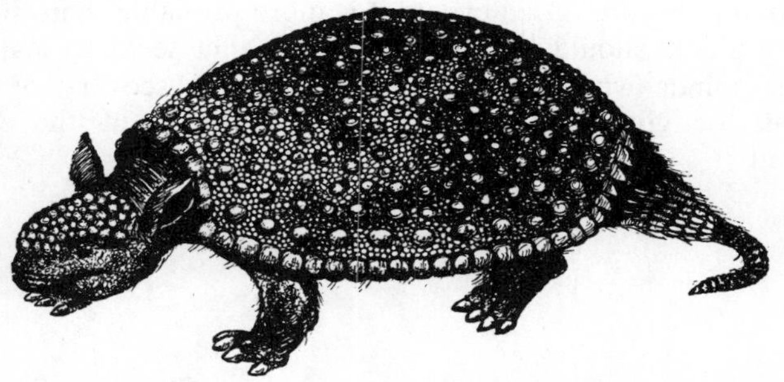

52. Reconstruction of the Glyptodon, a giant armadillo, 12 feet long.

armor. There is no reason why a giant armadillo should not take up a semiaquatic life. Like the giant dinosaurs, such as the Diplodocus, it might have taken to a medium denser than air, where its heavy body, laden with its great carapace, could move more easily. With its scales it could easily be mistaken for a giant snake if only its head and the huge dome, more than 3 feet wide, of its carapace were visible above water. The legends about gigantic reptiles might be due to people seeing Glyptodons and very large water boas and mixing the two together. More evidence is needed before one can draw such a bold conclusion, but given the comparatively recent date of the Glyptodons their survival would not be so surprising.

Personally I am inclined to think that a third animal may have confused the story. The monstrous water snakes are so similar to the great sea serpent—which is certainly not a snake, since it wriggles up and down, and not from side to side—that I wonder whether they may not be primitive Cetacea (Archaeoceti). As Remington Kellogg suggested in 1936 in his *Review of the Archaeoceti*, the first stages in the history of the Cetacea (whales, etc.), which are mammals that have reverted to the sea, may have occurred in fresh-water lakes and rivers. Some of the Archaeoceti had a coat of mail consisting of rings of horny scales around the body. The rows of nodules on the skin on some porpoises' backs (genus *Neomeris*) are a vestige of this ancient armor.

If a primitive cetacean had survived today, its head, rising above the surface on a neck as flexible as a seal's, and its scaly back would make it look very like a monstrous serpent.

But all this can be no more than a theory, for we know so little about the *minhocão*. At all events it is more probable than that the surviving giants should be those dinosaurs that seem to leap into romantic minds whenever they learn of the discovery of huge footprints or glimpses of vast aquatic animals in the Green Continent.

CHAPTER **12**

APES IN GREEN HELL

For mine owne part I saw them not, but I am resolued that so many people did not all combine or forethinke to make the report.

SIR WALTER RALEIGH on the tales of hairy men that he heard in South America.

THE FIRST REPORTS about the New World tell of hairy men. Yet the only monkeys known to science in America are relatively small and have long tails. During their discovery of the "Empire of Guiana" in 1595 and 1596 Sir Walter Raleigh and his continuator Laurence Keymis heard tell of them. And of course at this time, long before the idea of evolution was dreamed of, any animals whose characteristics were a mixture of those of two known animals—in this case man and the ape—were thought to be monstrous crossbreeds, as can be seen from the following passage from the Spanish historian Pedro de Cieza de Leon's *Crónica del Perú*:

> It is also said that in other places there are (though for myself I have not seen them) very large long-tailed monkeys which live in the trees and which the natives (tempted by the devil who seeks to find where and how he can make man commit the vilest sins) use like women and, it is affirmed, some of these monkeys are supposed to give birth to monsters which have a man's head and privy parts and a monkey's hands and feet. They have, it is said, a thin body, and a great stature. They are hairy. Indeed they resemble (if it is true they exist) the devil their father. It is also said that they have no language but a plaintive moan or howl.

Rumors of large apes, like the chimpanzee or the gorilla, have persistently come from South America, and sometimes significantly from remote places where they could not have been introduced from Europe. In 1769 Dr. Edward Bancroft brought from the heart of Guiana a report about an "orangutan"

> much larger than either the *African* [the chimpanzee] or *Oriental,* if the accounts of the natives may be relied on. . . . They are represented by the *Indians* as being near five feet in height, maintaining an erect position, and having a human form, thinly covered with short black hair; but I suspect that their height has been augmented by the fears of the *Indians,* who greatly dread them. . . .

When, some 30 years later, Alexander von Humboldt reached the still-unexplored rapids of the Upper Orinoco, he heard tell of "a hairy man of the woods" which was reputed to build huts, carry off women, and eat human flesh. The alleged habits of this lascivious beast remind one of those attributed to the gorilla, which, however, was still utterly unknown in Europe. According to Humboldt this fable could be based on "one of those large bears, the footsteps of which resemble those of a man and which are believed in every country to attack women." In 1860 Philip H. Gosse pertinently asked what "large bear" could inhabit Venezuela, for the only known South American bear is of very modest size. "Is not such a bear in South America quite as gratuitous as the monkey himself?" Rejecting Humboldt's explanation as unconvincing, Gosse affirms that it is possible that in South America "there may exist a large anthropoid ape, not yet recognised by zoologists." And today there is no lack of rumors about fantastic man-faced monsters in the Green Continent, much of which is still little known. Frank W. Lane writes:

> Gold prospectors, working on the River Araguaya, which flows near the Matto Grosso, have heard roars coming from the depths of the virgin forest. Cattle have been found dead and every time their tongues had been wrenched out by the roots. Two prospectors have seen footprints in soft sand by the river, which resembled those of a man, but were 21 inches long.

Henri Pitaud told me in 1955 that a similar event had caused a sensation in the Ybitimi region in Paraguay a few years before.

> In an *estancia* about a hundred cattle were found dead without any wounds except that the tongue was torn out and gone. This went on for nearly eight months. Everything returned to normal, then, sud-

denly, two or three years ago, the same events recurred in the same area but in a different *estancia*: another hundred cattle suffered the same fate.

The footprints are most puzzling, for though human-looking tracks may be made by an ape, a bear, or even a large reptile, in this area there is no known animal that leaves footprints 21 inches long. In fact they can only be attributed to certain prehistoric (or supposedly prehistoric) animals, such as the Megatherium or a dinosaur *Prestosuchus chiniquensis,* whose remains have been found in Brazil and whose footprints looked as if they had been made by enormous human hands.

But this could not explain the torn-out tongues, for surely it needs a true hand to grasp a cow's tongue, and extraordinary strength to succeed in ripping it out. Must it not *really* be some kind of ogre? The idea is so fantastic that it is hard to take seriously, but since there is now proof that more or less human-looking giants once inhabited a large part of the earth and may still survive in the Himalayas, it is impossible to banish this nightmarish notion.

It would indeed be odd if the vast basins of the Amazon and the Orinoco did not still hold several surprises for the zoologists, but so far most of the evidence about unknown animals there comes from isolated witnesses and for lack of confirmation is not accepted by serious naturalists. There is one great exception: the large ape, very human in appearance, which has been met several times and has even been photographed, though its identity is still in dispute.

In 1917 François de Loys, a Swiss geologist, set off with a handful of men deep into the Sierra de Perijaá, which lies astride the borders of Colombia and Venezuela and is inhabited by the dangerous Motilone Indians. The expedition went on for three years. It was an exhausted party of men, decimated by fever and skirmishes with the savages, who in 1920 were in the forests along the Tarra River, southwest of the lagoon of Maracaibo.

There, not far from the river, Loys and his companions suddenly met two tall monkeys which advanced toward them, walking upright and holding onto bushes. They seemed to be beside themselves with rage, screaming, waving, and tearing off branches and brandishing them like weapons. At last they reached such a pitch of fury that they defecated into their hands and hurled their excrement at the intruders. Loys and his companions aimed their guns at the male, who was in front and was the most threatening of the two, but he stepped aside and let the female pass. It was she who was killed by the salvo from the guns. Whereupon the male fled. The dead animal was then carried to the riverbank, seated on a fuel crate, and held upright by a stick under its chin. In this position it was photographed. (See Plate 12, following page 146.)

Professor Georges Montandon, to whom Loys reported the incident and who informed the learned world of it, points out that this monkey was "an utterly new apparition to the Creoles who accompanied the leaders of the expedition, though not perhaps to the Motilone savages." What was this large monkey? At first glance the only photograph we have of it reminds one irresistibly of a spider monkey, but with a strikingly human expression on its face—more human than that of any anthropoid ape. The nostrils are very wide apart, separated by a broad wall of cartilage: it is thus a Platyrrhinian, like all the other American monkeys.

53. Capuchin monkey, a typical Platyrrhinian monkey (with splayed nostrils) of the New World.

The thumbs of its hands are extremely small. The genitals (which are not a penis but a clitoris, for it is indeed a female) are strikingly large.

This anatomical peculiarity may be the origin of the South American legend that apes couple with women. Since all the spider monkeys seemed to be males, their mates would have to be of another species: Indian women.

Had Loys's monkey merely been a spider monkey—as many of its features suggest (to say nothing of the habit, frequent among these beasts, of hurling its excrement at its enemies)—he would hardly have gone to the trouble at such a critical moment of his explorations to photograph such a common animal. Besides, if he is to be believed, the monkey in question was far larger and more massive than any spider monkey he had ever seen, and it had no tail. All the spider monkeys have tails—and what is more they are long and prehensile.

This absence of tail, a feature in common with the African and Asian anthropoid apes, is, alas, not visible in the photograph. On the other hand, it is hard to deny that Loys's monkey has a more massive body and thicker limbs than the ordinary spider monkeys. A detail which I do not think has been pointed out is that its thorax seems to be flattened dorsoventrally, like an anthropoid's—as one can see from its broader shoulders—and it is much longer than a spider monkey's. Its face is also much more oval.

54. Spider monkey of Brazil.

There is one other very important characteristic in which the Platyrrhinian monkeys of the New World differ from the Catarrhinians of the Old World. The former have 36 teeth, the latter 32. Loys asserts that the animal he shot had only 32. We cannot verify this figure without examining the monkey's dentition in detail. Alas, the expedition's cook who prepared and preserved the animal's skull was rash enough to turn it into a salt box. It dried and disintegrated in the heat and little by little the pieces were lost.

If Loys has given the number of teeth correctly, the monkey could be a Platyrrhinian only if there were a freakish reduction in the basic number, which, as Dr. Charles Bennejeant has pointed out, is extremely rare: the American monkeys are more apt to increase the number of their molars. But Dr. Schultz of Baltimore, who collected some 400 specimens of monkeys in Nicaragua and Panama, has shown that the spider monkeys are an exception to this rule and often lack the third molar.

If Loys's ape is, as it seems, a highly developed spider monkey, one would expect that what is a tendency in the rest of the group would be a normal characteristic for it, in which case Loys was telling the truth, and we should indeed be all the more inclined to trust the rest of his account, especially about the absence of tail.

There remains its size. The largest American monkeys, howlers, spider monkeys, and woolly spider monkeys, are no more than 3 feet 7 inches high when they stand up on their hind legs. Loys swears his monkey was much larger. At first, relying on his memory, he gave a height of 4 feet 5 inches, but later he found a note he had sent home to his mother giving the exact dimensions: 1 meter and 57 centimeters—5 feet 1¾ inches. So large an American monkey would be an animal quite new to science (though it agrees exactly with the native tales reported by Dr. Bancroft), and the skeptics at once began to call Loys's assertions in question and to dispute his measurements. But the probable size of the fuel crate will determine the animal's size fairly accurately, as was demonstrated at the session of the Académie des Sciences at Paris on March 11, 1929, at which the "monkey of anthropoid appearance" described by Professor Montandon was presented.

According to M. Cintract, a photographer, who judged from the number of planks in the crate, it must have been about 20 inches high, and the animal's height was between 5 feet and 5 feet 3 inches. On the other hand the standard size for fuel crates is 17¾ inches high, and as the animal is three and a third times the height of the case, this implies a height of 5 feet.

Which confirms the accuracy of Loys's measurements. The monkey would therefore be taller than the chimpanzee, the female of which rarely exceeds 4 feet 3 inches in height. No wonder Loys was so surprised. On the strength of this photograph and information given him by the geologist, Professor Montandon published a careful description of the unknown animal in 1929, naming it *Ameranthropoides loysi,* to mark its anthropoid characteristics and honor its discoverer. Gosse's prediction had come true.

To his description of *Ameranthropoides loysi,* Dr. Montandon appended two curious documents which showed that large monkeys were not utterly unknown in America. The first was the passage from Pedro de Cieza de Leon which I have already quoted. The second is an account of stone statues like gorillas in the museum at Merida in Yucatan.

There are two of these creatures without legs, but standing upright more than five feet high on the stumps of their thighs . . . they were found near the town of Tekax in Yucatan. . . . One of these statues seems to be bisexual, for although it has male features, it carries a child on its left arm, like a mother. The figures have a strikingly apelike position. They have pronounced eyebrows, broad chests and a bent back.

Montandon goes on to make higher claims:

If our subject's well-developed forehead is not merely the result of an optical illusion, it is not impossible that this creature is a new genus of the Hominidae which would put it on a level with the genus *Pithecanthropus.*

The late Professor Leonce Joleaud said:

I see the new monkey discovered on the Rio Tarra taking its place in the world of the Cebidae and in particular of the spider monkeys, in a position determined by a stage of evolution comparable to that of the Pithecanthropus in relation to the group of gibbons.

It is now known that in the Sinanthropus, which was no doubt a variety of the Pithecanthropus, "the jaw was," in Professor C. Arambourg's words, "strongly prognathous, the nose platyrrhinian, and that altogether it had a pronounced pithecoid appearance." So it is hardly surprising that there should be such a striking resem-

55. Comparison of the shape and structure of the face of the Ameranthropoid (shown here with its mouth shut) and the Pithecanthropus (after an unduly humanized reconstruction by Von Koenigswald, 1938).

blance between the Ameranthropoid and a reconstruction of the Pithecanthropus made by Von Koenigswald in 1938, even though he has humanized it unduly. The bare skin, hair only on the head and eyebrows, lips, sclerotic visible in the eyes—all these are features which cannot be inferred from the bones alone, and must therefore have been invented. If you disregard them you will see that the shape of the face is the same in the two. Indeed it is very possible that the Pithecanthropus looked more like Loys's monkey than like the suspiciously humanized reconstruction of it.

On the other hand, the Ameranthropoid's face is clearly different from that of an ordinary spider monkey's. It is oval in outline instead of being rather triangular. The bottom of its face is much heavier, and its jaws are more powerful. And though it has been photographed full face, one can see that its profile is more like that of a capuchin monkey, another smaller American species, than like a spider monkey's, in which the upper jaw recedes sharply, giving it a fairly pointed snout, while the capuchin's is more rounded. No doubt Loys's monkey has a similarly slight prognathism.

56. Comparison of the profile of a spider monkey (*above*) and a capuchin (*below*).

It would be extremely interesting to know the Ameranthropoid's cranial capacity. If the development of its brain has reached the same level as the Java ape man's, its habits and psychology would be a fascinating study. The best-preserved Pithecanthropus skull (skull II) has a capacity of 775 cc., between the gorilla (600 cc. average) and a modern European (1,320 cc. average). The Sinanthropus, though so similar to the Java ape man, is better off, with an average of about 1,040 cc.

The cranial capacity is the only evidence from which the anthropologist can judge a mammal's cerebral development, and it has to be considered not absolutely (since an elephant's brain weighs four times as much as a man's and a whale's five times as much), but relatively to the weight of the body as a whole. Even so, a man's brain is only one forty-sixth of his total weight, whereas a capuchin's or a spider monkey's brain is between one fifteenth and one seventeenth. But as man is undeniably more intelligent than these monkeys, all sorts of formulas have been invented, with varying degrees of success, to calculate a "cephalization index" which will give man his pre-eminent place. The weight of the brain and the body are always the main factors, but many authors also take into account the height. Be that as it may, the cerebral development of the monkeys of the New World, as regards both weight and height, is much more remarkable than that of those of the Old World, even including the anthropoid apes. And as Bierens de Haan, the Dutch psychologist, has proved, the intelligence of the capuchin monkey, even though it is no bigger than a small cat, can very well be compared to that of the remarkably intelligent chimpanzee.

If, as in the African and Asian monkeys, the intelligence of the American monkeys is very roughly proportional to their size, the Ameranthropoid's intellectual faculties must be much higher than those of the chimpanzee or gorilla. Even if it is merely a giant spider monkey, its cranial capacity would be much larger than theirs. No doubt it is larger than the Pithecanthropus's and perhaps even than man's. If a spider monkey or a capuchin kept its usual proportions but were as large as a man it would obviously have a larger brain than his; and the Ameranthropoid (5 feet 1¾ inches) is the height of a human Pygmy (5 feet or less) or even a short man. Even though it certainly has an intellect far inferior to the most backward of men (otherwise it would have conquered the world!), a comparative study of its anatomy and physiology, especially that of its brain, with those of a man of the same size would be of immense interest and might enable us to establish for certain the relation between the weight, volume, and shape of the brain and the degree of intelligence.

The zoological problem of Loys's monkey demands a solution. Some eminent naturalists, like Sir Arthur Keith, have merely maintained that it is just a variety of ordinary spider monkey and that the tail has been hidden in the photograph—thus implying that Loys is a fraud. Quite apart from the size of the animal—which can still be disputed, though unfairly—it seems to me that the photograph proves that this is an unknown kind of monkey. But Philip Hershkovitz, an excellent American mammalogist who during the war prospected the area that Loys had explored but found no trace of the Ameranthropoid, even claims to have determined the species of spider monkey to which it belongs, the mulatto spider monkey (*Ateles hybridus*). I therefore spent a long time in the Natural History Museum in Paris studying the "type" of this species, but I was not in the least convinced. While it is true that the mulatto spider monkey has much thicker limbs than the other spider monkeys (at least it had in this specimen which had been mounted from a skin in poor condition) and fur not unlike the Ameranthropoid's, its size, the length of its thorax, and more especially the shape of its face quite refuted this identification.

Legends of large monkeys are not confined to the Colombo-Venezuelan border, but are found all over the country of the Amazon. The Marquis de Wavrin writes:

> They are called *maribundas*. Their height when standing upright, a position they readily adopt to walk on the ground, is supposed to be about 5 feet. The only civilized man who lives with his family on the Guaviare on the upper reaches of this river, told me he had brought up a young *maribunda* at his house. It was very friendly and amusing in all its pranks; finally its owner had to kill it because it did too much damage.
>
> The *maribunda*'s cry sounds strangely like a human call. On the Guaviare in particular I several times thought at first that it was Indians calling.

Are these *maribundas* the same as Loys's monkey? I think not. In fact *maribunda* or *marimonda* is the native name for what in English is called the marimonda spider monkey (*Ateles belzebuth*) —but this is never more than 3 feet 7 inches high. And Wavrin adds: "According to what the Indians told me, this monkey's body is rather slim. It likewise has a prehensile tail"—which does not agree at all with Loys's description.

At first sight Roger Courteville's evidence appears more helpful. He gives a most graphic account of an encounter with a large tailless ape. But, even if many of the details he gives were not suspect, how can he expect his story to be believed when he illustrates it with a photograph of his "Pithecanthropus" waving its arms among the branches, with the caption "Dr. de Barle's document"? This is obviously a crude fake. Loys's photograph has been cut up, its limbs rearranged into a threatening posture and the whole thing stuck on a background of virgin forest.

This suspect evidence must not be allowed to cast suspicion on Loys's account, which is supported by an indisputable *document* in the correct sense of the word.

In 1868, a century after Dr. Bancroft, Charles Barrington Brown heard new rumors on the Upper Mazaruni on the Venezuelan frontier that a sort of hairy men lived there.

> The first night after leaving Peaimah we heard a long, loud, and most melancholy whistle, proceeding from the direction of the depths of the forest, at which some of the men exclaimed, in an awed tone of voice, "The Didi." Two or three times the whistle was repeated, sounding like that made by a human being, beginning in a high key and dying slowly and gradually away in a low one. . . .
>
> The "Didi" is said by the Indians to be a short, thick set, and powerful wild man, whose body is covered with hair, and who lives in the forest. A belief in the existence of this fabulous creature is universal over the whole of British, Venezuelan and Brazilian Guiana. On the Demerara river, some years after this, I met a half-bred woodcutter, who related an encounter that he had with two Didi—a male and a female—in which he successfully resisted their attacks with his axe. In the fray, he stated that he was a good deal scratched.

In 1931 Professor Nello Beccari, an Italian anthropologist, Dr. Renzo Giglioli, and Ugo Ignesti, made an expedition to British Guiana, where one of their secondary objects was to attack the problem of Loys's ape. For in this area the fauna, flora, climate, and indeed the whole ecological pattern are the same as in the Sierra de Perijaá. On his return from several months in the interior, Beccari met the British Resident Magistrate, Mr. Haines, who was then living on the Rupununi. Haines told him that he had come upon a couple of *di-di* many years before when he was prospecting for gold. In 1910 he was going through the forest along the Konawaruk, a tributary which joins the Essequibo just above its junction with the Potaro, when he suddenly came upon two strange creatures, which stood up on their hind feet when they saw him. They had human features but were entirely covered with reddish-brown fur. Haines was unarmed and did not know what he could do if the encounter took a turn for the worse, but the two creatures retreated slowly and disappeared into the forest without once taking their eyes off him. When he had recovered from his surprise he realized that they were unknown apes and recalled the legend of the *di-di* which he had been told by the Indians with whom he had lived for many years.

Beccari was also told of other encounters with pairs of large apes. The Indians believed that the *di-di* lived in pairs and that it was extremely dangerous to kill one of them, for the other would inevitably revenge its mate by coming at night and strangling its murderer in his hammock. Beccari did not trust the more fanciful part of this story, but felt that it must have a kernel of truth. Loys had also met a pair, and so had Barrington Brown's woodcutter.

Most South American monkeys live in largish troops, and this habit alone suggests that this is a very peculiar species.

The only difference between the Perijaá ape and those described in Guiana is in the color of the fur, the first being grayish brown and the second reddish. "Assuming Haines and Loys both described the color with scrupulous exactitude," Professor Beccari observes, "we may think the Guiana ape is a distinct variety of the Venezuelan species, or even a different but closely related species." He adds that orangutans vary in color from a dark mahogany to a bright foxy red, though nobody yet knows whether they belong to different races or species, or whether they are merely individual variations.

In Brazil the tales of a large manlike ape sometimes become quite fantastic. They are not found in the north of the country near Venezuela and Guiana—the Ameranthropoid's alleged home—but in the southwest in the provinces of Amazonas, Matto Grosso, and Goyaz as well as Acre and Guaporé on the frontiers of Bolivia. The creature has various names: *mapinguary, pelobo,* and *pe de garrafa* or bottle-foot, for it leaves footprints that look as if they were made by sticking the bottom of a bottle in the ground. Its tracks are always regularly spaced and exactly aligned.

I need hardly say that the attempt to imagine what creature could leave such extraordinary footprints has produced some bizarre results. This is how it was described in about 1954 to Rui Prado Mendonça, Jr., a Brazilian hunter, by one of the oldest inhabitants of the Upper Araguaya:

> It is an animal of a fair height, distinctly human-looking, with long flowing hair on its head, and it has only one leg, with which it makes enormous leaps, always leaving a track of deep prints like the bottom of a bottle. Hence its name. As it has only one leg it cannot walk like other animals, but always stands erect. It is extremely savage and never crosses obstacles in its path. It always goes round them, and is therefore reputed to move in endless zig-zags. When it meets an enemy it fixes him with its eyes with such intensity that the victim is quite hypnotized and falls unresisting into its claws.

It is hardly surprising that this monster terrified even such a brave jaguar hunter as the old half-breed who told this tale.

There is not the slightest doubt that tracks which look as if they were made with the bottom of a bottle do exist. They have been seen by witnesses whose veracity is above suspicion, among them

Francisco Meirelles who pacified the Chavante Indians. He thought that the track attributed by the Chavantes to the *mapinguary* was made by a deer with a broken leg, though it is hard to see how a limping animal could have left such a regular track, or indeed how anything short of a race of three-legged deer could have accounted for all the tracks of the *pe de garrafa* seen in the Amazon jungle.

A hundred years ago a similar phenomenon in England gave rise to even more fantastic legends. On the morning of February 8, 1855, tracks very like a *pe de garrafa*'s were found in the snow around no less than 18 towns and villages in Devon. They were almost circular, or rather horseshoe-shaped, and in perfectly straight lines. But this beast did not seem to be afraid of obstacles, there were footprints everywhere, on roofs, the tops of walls, in walled gardens as well as open fields, and on both sides of the Exe estuary. The beast must have walked more than a hundred miles. Commander Gould wrote:

> A natural explanation of the facts seemed impossible to find, and difficult even to suggest; while any explanation certainly postulated the visit of something very uncanny—something which walked upon small hooved feet with a very short, mincing stride, which sought darkness and solitude, which had never rested, which had covered something like a hundred miles in a single night, which had crossed a river two miles wide, which had hung round human habitations without daring to enter them, and which had on some occasions walked up walls and along roofs, while at other times it had passed through such obstructions as if they did not exist.

There was a period of panic, during which people dared not go out after dark, and, as *The Times* wrote, "The superstitious go so far as to believe that they are the marks of Satan himself."

Far be it from me to suggest that both the *pe de garrafa* and the "Devil's hoof-marks" are due to the same hellish and ubiquitous beast, or to survivors of the Skiapods of antiquity, hiding in Brazil and Devon. But they may well have a common origin—a meteorological origin as Dr. Maurice Burton has suggested might be the cause of the "Devil's hoof-marks," though no meteorologist has been able to put forward a phenomenon that would explain them. There is no proof that the bottlelike marks in Brazil have anything to do with the tales of ape men current in the area. It would not be the first time that two sets of facts have been wrongly related, just because they were both mysterious. Admittedly an arboreal ape walks on the outside edge of the foot, thus leaving a more or less ring-shaped mark. An orangutan's footprint tends in this direction (see page 68). But, of course, this could not explain the Devonshire abominable snowman.

The fact that the footprints were in a dead straight line does not necessarily mean that the beast that made them had only one leg. It could have been a biped. The few people who have seen a *mapinguary* describe it much more prosaically than the old half-caste from the Upper Araguaya, as we shall see from an account told to the Brazilian writer Paulo Saldanha Sobrino, by a half-caste called Inocêncio, who in 1930 went on an expedition up the Uatumã toward the sources of the Urubú. When their boat came to an impassable waterfall they cut across the jungle to reach the Urubú watershed. After two days Inocêncio became separated from the rest of his party. He shouted and fired his gun, but there was no reply except the chatter of monkeys and squawks of angry birds. So he began to walk almost blindly, feeling he must do something in such a critical situation, until night fell, when he climbed into a large tree and settled himself in a fork between the branches. As it grew dark the night was filled with jungle noises, and Inocêncio rested happily enough until suddenly there was a cry which at first he thought was a man calling, but he realized at once that no one would look for him in the middle of the night. Then he heard the cry nearer at hand and more clearly. It was a wild and dismal sound. Inocêncio, very frightened, settled himself more firmly into the tree and loaded his gun. Then the cry rang out a third time and now that it was so close it sounded horrible, deafening and inhuman.

> Some forty yards away was a small clearing where a *samaumeira* had fallen and its branches had brought down other smaller trees. This was where the last cry had come from. Immediately afterward there was a loud noise of footsteps, as if a large animal was coming toward me at top speed. When it reached the fallen tree it gave a grunt and stopped. . . . Finally a silhouette the size of a man of middle height appeared in the clearing.

The night was clear. There was no moon, but the starry sky gave a pale light which somehow filtered through the tangled vegetation. In this half-light Inocêncio saw a thickset black figure "which stood upright like a man."

> It remained where it stood, looking perhaps suspiciously at the place where I was. Then it roared again as before. I could wait no

longer and fired without even troubling to take proper aim. There was a savage roar and then a noise of crashing bushes. I was alarmed to see the animal rush growling toward me and I fired a second bullet. The terrifying creature was hit and gave an incredibly swift leap and hid near the old *samaumeira*. From behind this barricade it gave threatening growls so fiercely that the tree to which I was clinging seemed to shake. I had previously been on jaguar hunts and taken an active part in them, and I know how savage this cat is when it is run down and at bay. But the roars of the animal that attacked me that night were more terrible and deafening than a jaguar's.

I loaded my gun again and fearing another attack, fired in the direction of the roaring. The black shape roared again more loudly, but retreated and disappeared into the depths of the forest. From time to time I could still hear its growl of pain until at last it ceased.

Dawn was just breaking.

Not until the sun was well up did Inocêncio dare to come down from his perch. In the clearing he found blood, broken boughs of bushes, and smashed shrubs. Everywhere there was a sour penetrating smell. Naturally he did not dare to follow the trail of blood for fear of meeting a creature which would be even more dangerous now that it was wounded. Taking a bearing on the sun he at last reached a stream and rejoined his companions, who fired shots so that he should know where they were.

I maintain I have seen the *mapinguary* [Inocêncio said to Paulo Sobrino]. It is not armored as people would have you believe. They say that to wound it fatally you must hit the one vulnerable spot: the middle of the belly. I can't say where it was wounded by my bullet, but I know it was hit, for there was blood everywhere.

This story has the ring of truth, and is told in more sober fashion than most Brazilian hunters' tales, true though they often are. Inocêncio claims to have done nothing to boast of, and he does not make the animal in the least fantastic. It has two legs like everybody else. Like the legendary *pe de garrafa* it emits terrible roars, but they do not send the hearer mad, and likewise it gives off a strong smell, but it is not really asphyxiating. Its behavior is just what one might expect of a powerful great ape—a great ape like the one shot by Loys in Venezuela and seen by Haines and others in Guiana—the only unknown animal of which we have an excellent photograph, and whose existence cannot, I think, be disputed except by the disingenuous and the blind.

PART FIVE

THE GIANTS OF THE FAR NORTH

The creation of the mammoth was a blunder of the Superior Being. In creating such an enormous animal, the Creator did not take into consideration the size of the earth and its resources. One earth could not stand the weight of the mammoth and its vegetation was not sufficient to feed the mammoth race. The mammoth fed on tree trunks which he ground with his teeth, and in a short time the whole North of Siberia was deprived of trees. Hence is the origin of the northern tundra. In the beginning the earth had the form of an even plain, but by his weight the monster animal in moving about caused the formation of valleys and ravines in which rivers originated. In swampy or sandy places the mammoth sank into the ground and disappeared under the earth, where he froze during the winter. Often in the hole over him water gathered into a lake. In this way the mammoth gradually disappeared from the earth's surface. This is why now whole cadavers of the animal are to be found in the frozen soil.

Yukaghir tradition quoted by WALDEMAR JOCHELSON.

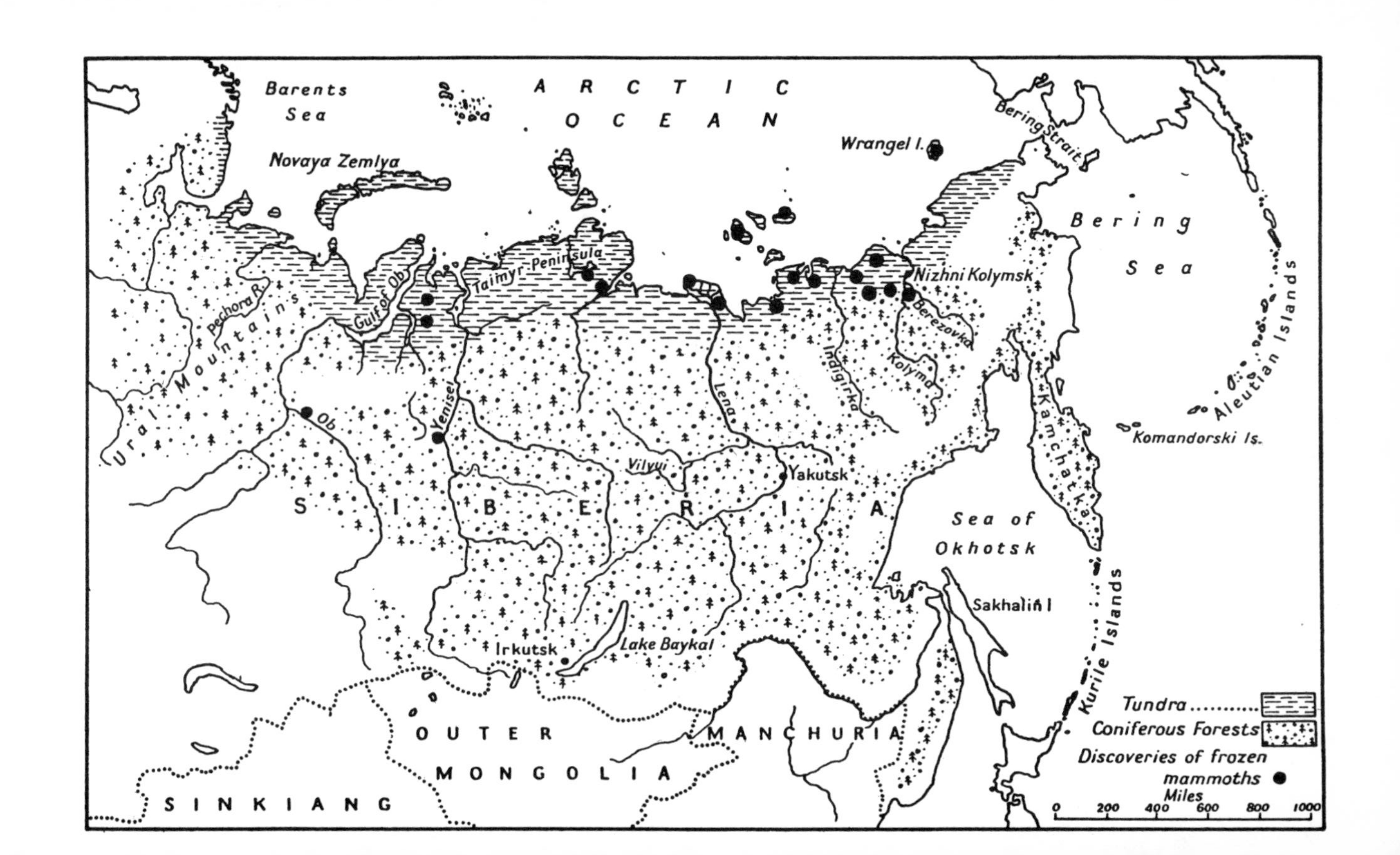
ARCTIC
OCEAN
Barents
Sea
Novaya Zemlya
Wrangel I.
Bering Strait
Bering
Sea
Aleutian Islands
Komandorski Is.
Nizhni Kolymsk
Berezovka
Kolyma
Indigirka
Lena
Taimyr Peninsula
Gulf of Ob
Pechora R.
Ural Mountains
Ob
Yenisei
Vilyui
Yakutsk
SIBERIA
Kamchatka
Sea of
Okhotsk
Sakhalin I
Kurile Islands
Irkutsk
Lake Baykal
OUTER
MONGOLIA
MANCHURIA
SINKIANG
Tundra
Coniferous Forests
Discoveries of frozen
mammoths
Miles
0
200
400
600
800
1000

CHAPTER **13**

THE MAMMOTH OF THE TAIGA

This Creature, though rare, is found in the East of the Northern Siberian Zone.

HILAIRE BELLOC, *The Frozen Mammoth.*

THE LAPPS BELIEVE in monstrous beasts which live under eternal snows; and throughout northern Siberia we find the same belief in a hairy monster of underground habits. The Chuklukmiut Eskimos on the western shore of the Bering Strait call it *kilu kpuk*. The Chukchi, who live in the north of the eastern tip of Asia beyond the Kolyma River, think that a similar monster lives beneath the ground and moves along narrow passages. When a man sees the monster's tusk sticking out of the ground he must dig it up at once, otherwise it disappears again and carries curses in its train elsewhere. Once some Chukchi saw two tusks coming out of the earth. They began their incantations and the beast miraculously appeared and provided them with a winter's supply of meat. From their descriptions there is no doubt that the monster is the mammoth known to paleontologists.

The Yukaghir, whose territory stretches beyond the Arctic Circle all along the Arctic Ocean from the Lena delta to beyond the Kolyma, also speak of the mammoth in their traditions and call it *xólhut*. They even have a traditional memory of the mammoth's disappearance, which I have quoted as an epigraph to this part. Oddly enough many scientists today explain its extinction in the same way. Among the Yukaghir of Nizhni Kolymsk who have come under Russian influence, the story has altered and taken on a Biblical flavor. Noah, they say, wished to take a pair of mammoths aboard the Ark, but when the first of them put its feet on board, the ship almost capsized. So Noah hurriedly put to sea, leaving the mammoths behind, and thus they all perished.

Farther south, both among the Yakuts and the Ostiaks, and also among the Koriaks, there are very similar legends about a sort of giant rat called *mamantu* or "that-which-lives-beneath-the-ground," from which the word "mammoth" is derived. It cannot survive the light of day. The moment it comes into the light it is struck dead.

In Mongolia and Manchuria the mammoth's ghost appears as a gigantic mole. It is a hairy monster with tiny eyes and a short tail; and it digs enormous tunnels in the snow with its two teeth shaped like picks. In *The World of Animals,* attributed to the Emperor K'ang-Hi (1662–1723), there is also an allusion to the Siberian legends:

> One can read in the ancient book *Chin-I-King* about the *fen-shü* . . . There is in the North in the country of the Olosses near the sea, a kind of rat as big as an elephant which lives underground and dies as soon as it comes into the air or is reached by the sunlight.

The imperial author adds some more matter-of-fact details:

> There are *fen-shü* which weigh as much as 10,000 pounds. Their teeth are like those of elephants: the natives of the North make bowls, combs and knife-handles, etc. out of them. I have myself seen these teeth and these tools made from them, and so I believe in the truth of our ancient books.

The Mirror of the Manchu Language (1771 edition) gives curiously exact information about the mammoth:

> The rat of the ice or of mountain streams lives in the earth in the countries of the North, under thick ice. One can eat its flesh. Its hair is several feet long, and is used for weaving cloth upon which damp mists condense.

The natives of Siberia support their stories with the tangible evidence of ivory tusks, some of them as much as 16 feet long and weighing 450 pounds. There was once a flourishing trade in this ivory. At the beginning of this century the province of Yakutsk exported an average of 152 pairs of tusks per annum. During the last 200 years the tusks of some 25,400 beasts must have been sent out of this province alone. The trade is very ancient: it was mentioned in Chinese chronicles before the Christian era. Pliny the Elder had heard of it from the writings of Theophrastes, a pupil and successor to Aristotle, who reported that ivory was extracted from the ground.

In 1611 an English traveler called Josias Logan exhibited in London an elephant's tusk that he had brought back from Russia. Everyone knew that elephants were only found in Africa and India. It was true that their remains had been dug up in Europe, but these were thought to have been left by Hannibal's armies.

In 1692 a Dutch diplomat called Evert Ysbrants Ides went to China on Peter the Great's behalf to make a peace treaty with the Emperor K'ang-Hi. He reported:

> Amongst the Hills which are situate North-East of, and not far from hence [the village of *Makofskoy* near *Jenizeskoy*], the *Mammuts Tongues and Legs are found;* as they are also particularly on the Shoars of the Rivers *Jenize, Trugan, Mongamsea, Lena,* and near *Jakutskoi,* to as far as the *Frozen Sea.* In the Spring when the Ice of this River breaks, it is driven in such vast quantities, and with such force by the high swollen Waters, that it frequently carries very high Banks before it, and breaks off the tops of Hills, which falling down, discover these Animals whole, or their Teeth only, almost frozen to the Earth, which thaw by degrees. I had a Person with me to *China,* who annually went out in search of these Bones; he told me as a certain truth, that he and his Companions found a Head of one of these Animals, which was discovered by the fall of such a frozen piece of Earth. As soon as he opened it he found the greatest part of the Flesh rotten, but it was not without difficulty that they broke out his Teeth, which were placed before his Mouth as those of the Elephants are; they also took some Bones out of his head, and afterwards came to his Fore-foot, which they cut off, and carried part of it to the City of *Trugan,* the Circumference being as large as that of the wast of an ordinary Man. The Bones of the Head appeared somewhat red, as tho' they were tinctured wth Blood.

The Yakuts, Tungus, and Ostiaks told him the usual legends about a subterranean monster:

> But the old *Siberian Russians* affirm that the *Mammuth* is very like the *Elephant;* with this only difference, that the Teeth of the former are firmer, and not so straight as those of the latter. They also are of the Opinion, that there were Elephants in this Country before the Deluge, when this Climate was warmer, and that their drowned bodies floating on the surface of the Water of that Flood, were at last wash'd and forced into Subterranean Cavities: But that after this *Noachian* Deluge, the Air which was before warm was changed to cold, and that these Bones have lain frozen in the Earth

ever since, and so are preserved from putrefaction till they thaw, and come to light; which is no very unreasonable conjecture.

This very reasonable conjecture seems to have been overlooked, while the absurd tales of elephants that behaved like moles discredited these reports, and many European naturalists gradually came to the conclusion that they probably referred to an animal only distantly connected with the elephant, but also having large tusks of fine ivory: the walrus (*Odobaenus rosmarus*). Remains of this huge pinniped, which may reach a length of 16 feet and have tusks 2 feet long, had no doubt been confused with an elephant's. The tusks were of course much smaller, but the animal did at least live on the Siberian coast.

57. Head of a male walrus.

While the scientists in Europe were puzzling over the problem to no purpose, Peter the Great, whose curiosity had been aroused by Ides's reports, sent a German naturalist, Dr. D. G. Messerschmidt, of whom he had a particularly high opinion, to explore Siberia and try to solve the question of the incredible burrowing elephants. Either by good luck, or because mammoths' carcasses were extraordinarily common, Messerschmidt happened to hear that one of these beasts had just emerged from the melting ice of the Indigirka. In 1724 a Russian soldier had arrived just in time to examine the remains, already largely putrefied and eaten by wolves. There was little left but the head. However, another witness called Michael Wolochowicz had also seen a huge strip of rotten skin which was sticking out of a sand dune there. It was thick and brown and covered with hairs like a goat's.

Then, in 1771 the great German explorer and naturalist Peter Simon Pallas found among the melting glaciers of the Vilyui, a tributary of the Lena, part of a carcass still covered with thick dark-brown fur, mixed with long black hair like horsehair. He thought it must be a mammoth, yet there was no doubt that it was a rhinoceros. He therefore came to the conclusion that this beast was the origin of the legendary mammoth. For who would have expected that Siberia should yield not only hairy elephants, but also woolly rhinoceroses?

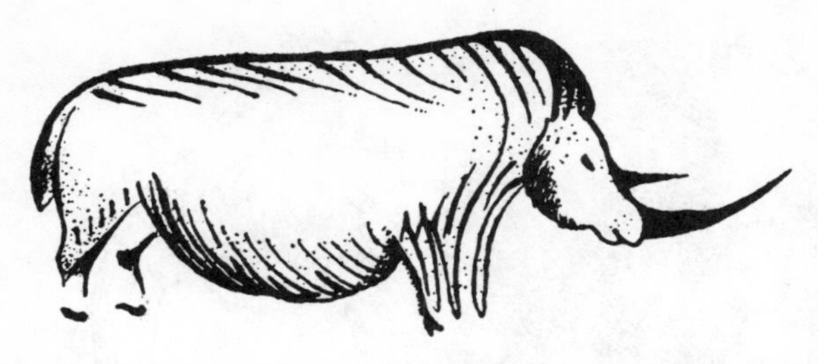

58. Woolly rhinoceros drawn on the side of a cave at Font-de-Gaume in the Dordogne (after the Abbé Breuil).

In the Middle Ages when fossil remains of mastodons, mammoths, or rhinoceroses were accidentally dug up in Europe they were thought to be relics of giant men. Bones of these animals were preserved in churches and attributed to dragons. And the mammoths' tusks were thought to be unicorns' horns. Now, in 1799, Johann Friedrich Blumenbach, of Göttingen University, solemnly announced on the evidence of the bones he had collected that an elephant had once lived in Europe, and that its tusks were curiously curved in an arc of a circle, quite unlike those of living elephants. He named it *Elephas primigenius*—not very aptly, since it was more specialized than modern elephants, and had been preceded in earlier ages by many other trunked animals. Naturally he did not identify it with the huge hairy mole that the Siberians called *mammont*.

While this was going on in the West a mammoth appeared in flesh and blood to a Tungus chief called Ossip Shumakhoff. The beast was embedded in a block of melting ice somewhere in the Lena delta and only a distorted picture of it was visible through its glassy case. Ossip fled in superstitious terror, but curiosity overcame his fear and he came back from time to time to see if it was still there. Two years after he had first seen it, one of the tusks emerged from the ice, and he was much tempted by the value of the ivory, but still terrified, for he had heard tales of people who had died from the mere sight of the monster. Fortunately a Russian merchant called Boltunoff saw a chance of a profitable bit of business and managed to make Ossip forget his fears and take him to the mammoth. In 1804 they went and found the beast now quite free of its shell; at first Ossip nearly died of fear, but soon he was helping Boltunoff to remove the tusks, and the shrewd Russian merchant immediately bought them for the miserable sum of 50 rubles.

Boltunoff also made a crude sketch of the beast which eventually came into Blumenbach's hands. It did not look much like an elephant—its trunk had no doubt been ripped off by hungry wolves;

59. The mammoth as drawn by Boltunoff.

the pressure of the ice had strangely twisted its tusks; its ears had come off, and as a result of decomposition Boltunoff had mistaken the earholes for eye sockets! But Blumenbach had no difficulty in recognizing his *Elephas primigenius*. Boltunoff's sketch of one of the molars left no doubt that it was the mammoth, and that the legendary beast was neither mole nor rhinoceros.

Meanwhile a Russian botanist called Professor Adams had heard that a mammoth had been found preserved whole, and set off to see for himself. When he arrived he found wolves, arctic foxes, wolverines, and even the Yakuts, had all been there before him. Apart from the skeleton, from which one foot was missing, there was not much left of the beast: an ear covered with silky hair, an eye, the brain, some tendons, and the lower part of the legs. Three quarters of the skin had been preserved. Adams set about collecting the relics. His native guides helped him to remove the skin which was almost an inch thick in places and needed 10 men to lift it. Then they carefully swept the ground and collected 37 pounds of hair. Everything was packed up with infinite care and sent to St. Petersburg, where the skeleton was scrupulously remounted. The curators of the "Cabinet of Rarities" founded by Peter the Great had bought it from Professor Adams for 8,000 rubles—160 times as much as poor Ossip got for his precious tusks. It may still be seen in the Zoological Museum at Leningrad.

At least a score more frozen mammoths were discovered in northern Siberia, the most recent being in 1935 and 1948, and there is no reason why many more should not come to light.

In April 1901 the Imperial Academy of Science of St. Petersburg learned that a mammoth in a perfect state of preservation was imprisoned in a melting glacier on the banks of the Berezovka. The Academy hastened to send off an expedition under Dr. Otto Herz, an entomologist whom one might expect to chase nothing larger than butterflies. This time wolves and other scavengers had not been able to eat all the carcass. Only the head emerged from the ice and had begun to thaw; and only the trunk, the chief delicacy, had been eaten. But bacteria were already hard at work, and an appalling stench arose from the putrefying meat.

To cut up a carcass which would gradually turn into an ever fouler cesspit was certainly a task to make any but the most devoted scientist blench. But the expedition included a young German taxidermist, E. W. Pfizenmayer, who had always dreamed of excavating prehistoric monsters. He now found his wish coming true with a vengeance. He was able to dissect the animal and study the smallest details of its anatomy. As he had several more opportunities again later, he eventually became the world's greatest authority on the mammoth.

As a result we now have an excellent idea both of the habits and exact appearance of the Siberian mammoth (*Elephas berezovkius*). It was covered all over its body with a reddish-yellow woolly fleece and also with long black hairs like horsehair, between 1 foot and 2 feet 3 inches in length, which hung from its cheeks, lower jaw, shoulders, flanks, and abdomen, making a sort of overcoat which reached almost to the ground. The color of the wool has faded with time and the beast was probably mainly reddish black. It had a short tail, under the root of which was a flap of skin which protected its anus from the cold.* Its thick skin was lined with a layer of fat about 3½ inches thick, and on the crown of its head and on its withers it had two fatty humps, which, like the camel's, served as a reserve of food.

Its concave forehead and small ears remind one of the Indian elephant (*Elephas indicus*). Between the frozen animal's stomach and diaphragm a large mass of clotted blood was found. This was carefully preserved and subsequent serological reactions proved conclusively that the Siberian mammoth was closely related to the Indian elephant.

Here I should say a word about the size of the mammoth. In

* The Indian elephant has a similar flap which probably acts as a protection against insects. This common feature shows the close relationship between the two species.

Le Petit Larousse Illustré you will find under *Mammouth* a picture of a man looking at a reconstruction of the hairy elephant. From their relative sizes the animal seems to be about 20 feet high at the withers. No mammoth was ever so enormous. True, there are various species of mammoth; there was once a giant North American species (*Elephas imperator*) which was a little over 13 feet high. Fraas's German mammoth (*E. fraasi*) was no doubt some 18 inches taller, but the Siberian mammoth was not even 10 feet high. It was a little smaller than the Indian elephant, which sometimes reaches that height, while the African elephant may be as much as 12 feet high at the withers. All the same the Siberian mammoth's humps and its shaggy coat must have made it an impressive mountain of fur.

We also know exactly what the Siberian mammoth ate. Thanks to Pfizenmayer, who in his first dissection removed from the stomach over 30 pounds of food which had been chewed but not digested, botanists were able to reconstruct its diet: the menu began with a species of gentian, its main dish consisted of grasses

60. The mammoth as it really was.

and sedges seasoned with wild thyme, for dessert there were two kinds of mosses, all flavored with alpine poppy, upright crowfoot, and a sort of northern orchid. The seeds present showed that the beast had died in the autumn. As Pfizenmayer pointed out:

> The mammoth's food is composed of the same plants that still grow today in the close neighborhood of the site of the discovery, plants that we have gathered and preserved in order to compare them.

All this might have served as a good argument that mammoths might still exist, had not modern science strangled these legends at birth. Cuvier's Theory of the Revolutions of the Globe, by which it was a priori impossible for any creatures of a past age to have survived, convinced most scientists in the beginning of the nineteenth century that the mammoth, whose fossil remains had been found almost everywhere, was a vanished species. The specimens found intact frozen in glaciers *must* therefore have been preserved in deepfreeze for 10,000 to 100,000 years according to various estimates. Nevertheless, John Frere had shown in 1797 that man was contemporary with large extinct animals when he found in Sussex worked flint weapons mixed with their bones. In 1823 the French geologist Ami Boué sent Cuvier a human skeleton excavated from ancient deposits which also contained the remains of extinct animals. Cuvier hushed up the awkward discovery. Similar discoveries soon became ever more frequent throughout the nineteenth century. Yet in 1863, Élie de Beaumont, Permanent Secretary of the Académie des Sciences, was still able to state categorically: "I do not think the human species was contemporary with *Elephas primigenius*. Cuvier's opinion was the work of a genius, it has not been overthrown."

A year later, Édouard Lartet, the founder of human paleontology, whose fundamental thesis the Académie had refused to publish, discovered at La Madeleine a flat piece of ivory upon which a Stone Age man had engraved a masterly drawing of a mammoth. Naturally this was thought to be a fake. It clearly shows the fatty hump on the skull, which at that time no one knew anything about. Not until the Berezovka mammoth was dissected in 1901 was it conclusively proved to be genuine.

Two points were now established: first that the mammoth, whose remains had been found all over Europe from the Arctic to the Alps and the Pyrenees, once spread over a vast territory; second that the strikingly accurate drawings of it in caves in the Dordogne

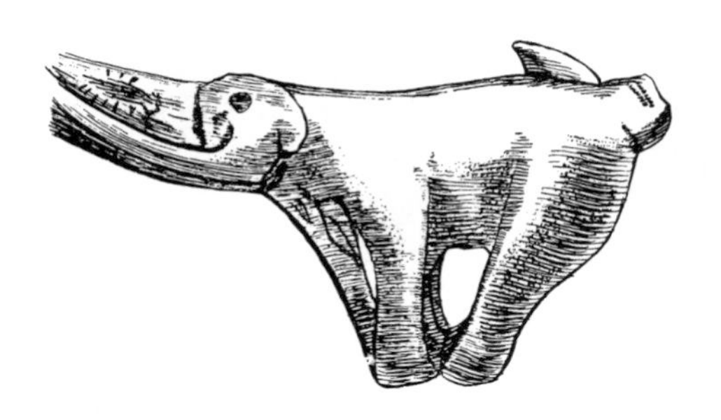

61. Mammoth carved in reindeer horn found at Bruniquel in France in 1868.

showed that it had been contemporary with man. As the mammoth had vanished, but man survived, Cuvier's notion of a revolution of the globe, or universal cataclysm that spared hardly anyone, was no longer tenable. Lyell had shown that there was no universal deluge or cataclysm. So what had caused the mammoth's extinction?

Mammoths had certainly survived in northern Siberia much longer than in central and western Europe—and for obvious climatic reasons. Any animal with thick fur is plainly adapted for a cold climate. The mammoth's thick coat shows that it belongs to a cold region. If it once lived over most of Europe and North America, the temperature was then much colder there than now. And we now know that glaciers formerly spread for a long period over the whole of northern Europe, transforming what had been

62. Picture of a mammoth on the wall of the cave at Les Eyzies in the Dordogne (after Capitan, Breuil, and Peyrony).

a world of lush vegetation inhabited by animals now found only in the tropics into a Siberian tundra, and that since the end of the Tertiary era this icecap has advanced and retreated four times.

In Europe the mammoth (*Elephas primigenius*) and the rhinoceros (*Rhinoceros tichorhinus*) belong to the last glacial period, which took place not more than about 12,000 years ago. It seems logical to suppose that these hairy elephants and woolly rhinoceroses followed the retreating glaciers northward when the temperature grew too warm for their liking after the long winter, which had lasted more than 50,000 years. But if they deserted the countries that had become too warm as thick forest gradually spread over the ancient steppes and even the tundra, this does not explain why they vanished from northern Siberia; for there they could still have found what seems to be their favorite climate.

63. Mammoth in the cave at Font-de-Gaume (after the Abbé Breuil).

At the end of the last century one of the greatest experts on the mammoth, Frederic Lucas, had to admit that he could not explain the extinction of an animal which was not only well protected against the cold, but also seemed actually to prefer low temperatures.

In 1919 this unsolved problem led Neuville to wonder whether the mammoths had not become extinct because of the cold rather than in spite of it. After all, they must have suffered a very low temperature for their carcasses to remain preserved until today. In their stomachs had been found remains of plants which are not characteristic of a severe climate; they are typical meadow flora.

Does this mean that a sudden wave of cold had destroyed the vegetation and starved the mammoths? Neuville thinks not, since such a sudden cataclysm could not account for their extinction over so wide an area. And if this had happened their stomachs would have been found empty, their reserves of fat exhausted and their bodies wasted away.

Their extinction does not seem attributable to any outside cause. Neuville found that their skin, like that of living elephants, had no sebaceous or sweat glands. Without them, without a constant flow of sebum and sweat to oil the hair and keep the fleece waterproof, the mammoth's thick fur would be no protection against cold and damp. Snow and icy rain would be able to penetrate its thick coat, soaking it through and through until it became a wretched cloak of ice, frozen to its skin. Neuville also pointed out that the exaggerated curve of the tusks, which sometimes form three quarters of a circle, made them useless weapons, indeed nothing more than rather cumbrous ornaments. In short, its extinction was due to progressive degeneration, due to its lack of adaptation to cold, aggravated by other causes and perhaps accelerated by an increasing lack of food.

Neuville's arguments do not seem to me very conclusive. The permeability of the mammoth's fur is largely compensated for by the unusually thick layer of fat beneath its skin. Moreover, the flap of skin covering its anus shows that its smaller details were well adapted to very low temperatures. And obviously if it was so ill-protected against the cold it would have stayed in Europe after the glaciers retreated. It would not have fled from the heat of the south to die of cold in Siberia. Even the argument about its excessively curved tusks is not very convincing. The elephant uses its tusks as clubs as well as spears; it does not usually transfix it enemies, it crushes them. A mammoth's tusks are not therefore such negligible weapons.

The latest and most generally accepted theory blames the mammoth's extinction on a series of accidents. When the hordes of mammoths followed the last of the retreating glaciers they were gradually led to the northeast. When they inadvertently ventured on to the marshy plains of North Siberia, their own weight made them sink into the icy mud, where they have been preserved ever since. This explanation certainly agrees curiously well with that given by the Yukaghir traditions, quoted as an epigraph on page 197.

That such catastrophes were often repeated is more than likely, but is it right to explain the *total* extinction of a species by a series of accidents? Animals do not leap into nature's traps with their eyes shut. It is hardly likely that all the mammoths in the world would have chosen to live in the one place where they had the greatest chance of being buried alive.

Perhaps we first of all ought to prove conclusively that the mammoth really is extinct. Many other animals that lived in western Europe with the mammoth during the glacial period left it, but they all found refuge in other countries, such as Alaska and Siberia. Why not the mammoth? Could it not have survived until today in a land as little known as the country of the Yakuts, the Ostiaks, and the Yukaghirs, where legends of "hairy giants" have always been current?

It seems to me that these legends, which may be thought to be a strong argument in favor of the mammoth's survival, are more likely to be due to frozen specimens in a perfect state of preservation. One of them was enshrined in a block of ice, standing up and facing the sea. It began to thaw, the beast was already partly free of the glacier, it seemed to be rearing up with its huge tusks raised in a threatening manner. The man who saw it was struck with terror and was said to have gone on for a long time about "the evil light in the giant's eyes."

Simple people, seeing one from afar, would not dream that it could be dead, let alone dead for thousands of years. This explains the legend about the giant mole living in burrows under the snow, which seem so out of keeping with an elephant; it refers to the fact that they are always seen rising out of the snow, and they die as soon as they are exposed to the light, since they are never seen to move. The discovery of giant footprints and huge excrements can be explained by the thawing out of tracks where mammoths once passed.

Here one might close the question of the living mammoth, were it not for something which everyone has taken for granted without inquiring whether it is true: the assumption that *the mammoth was an animal of the tundra and the arctic prairies*. Admittedly it is almost always in the tundra that the carcasses of frozen mammoths have been found. But is this a sufficient reason? Obviously they are most likely to be frozen whole in these icy and marshy plains; but might they not prefer to live elsewhere? For instance the fact that fossil remains of monkeys, lions, and hippopotamuses have been found in Europe does not mean that these animals belong to temperate regions, but that our countries once had a warmer climate.

The present tundra is an exactly similar case. The temperature in the whole of Europe was once much higher than it is today, particularly between the second and third glacial periods. The higher temperature was especially marked in Asia, where the glaciers did not come. Therefore it is likely that the taiga, the vast forest which covers all the middle of Siberia, once stretched farther north and covered the very place where most of the remains of mammoths were found. Must not the temperature have been milder to lead such animals as elephants and rhinoceroses to venture into these regions?

The analysis of the Berezovka mammoth's stomach contents shows that at the end of the summer it ate typical meadow flora. But subsequent analyses of other mammoths have proved that in winter the mammoth browsed on the leaves and branches of trees. It therefore seems that in winter at least the mammoth did as the reindeer does today and went into the taiga, and only in the spring and arctic summer did it venture out into the plains, which were suddenly covered with tasty grasses and scented flowers.

Although the mammoth's skin is a poor protection against the wet, it is not so against the cold. And there is one place where a beast of the mammoth's size could shelter from rain and snow: in the forest under trees. Certainly not on the bare and icy plains.

The theory that the mammoth belongs to the forests was confirmed after the original edition of this book was published, by the recent work of the Soviet academician E. N. Pavlovsky on a mammoth's carcass, excavated in 1948 from the banks of the Mamontova River. Vegetation dating from the same period was found beside the animal, and showed that the climate was less severe when the mammoth perished.

Professor Raymond Vaufrey concludes:

> The mammoth is not an arctic mammal; the climate in the areas where its frozen carcass is found was warmer than today; the forest zone reached much further north. In the spring and summer it is probable that the mammoths migrated to the north . . .

When the mammoths left western Europe they must have passed through at least 600 miles and sometimes more than 2,500 miles of taiga to reach the places where their frozen bodies were found. In this endless forest of pine and birch, broken only by lakes and marshes, they would have found plenty of food, a choice of drinking places, and perfect shelter under the scattered trees, where they could graze in peace and move their bulky bodies with ease under the protecting branches.

There is, of course, no need to assume that hordes of mammoths had carried out a vast organized retreat from Moscow in reverse. No doubt they once lived wherever there was a moderately cold climate and great pine forests like the taiga. They died out wherever the forest became too warm or thinned out so much that it was no longer any protection. In which case is it not possible that they may survive in the Siberian taiga?

The taiga is the largest forest in the world: it covers nearly three million square miles, more than thirty times the area of Great Britain, and about three quarters of the area of the United States of America. Herds of hundreds or even thousands of mammoths could easily live there without running the least risk of being seen by man. In fact the extraordinary thing is that some of these mammoths sometimes *are* reported to have been seen by the Ostiaks and Yakuts who live in the forests.

There is also a more circumstantial account of an adventure which is supposed to have occurred to a Russian hunter in 1918. He told his story in 1920 to L. Gallon, who was then *chargé d'affaires* at the French Consulate at Vladivostok. The hunter was a tall, elderly man, with very bright eyes, white beard and hair, and a tanned face seamed with scars. The French diplomat invited him to lunch, during which he told the following extraordinary tale:

> The second year that I was exploring the taiga, I was very much struck to notice tracks of a huge animal, I say huge tracks, for they were a long way larger than any of those I had often seen of animals I knew well. It was autumn. There had been a few big snowstorms, followed by heavy rain. It wasn't freezing yet, the snow had melted, and there were thick layers of mud in the clearings. It was in one of these big clearings, partly taken up by a lake, that I was staggered to see a huge footprint pressed deep into the mud. It must have been about 2 feet across the widest part and about 18 inches the other way, that's to say the spoor wasn't round but oval. There were four tracks, the tracks of four feet, the first two about 12 feet from the second pair, which were a little bigger in size. Then the track suddenly turned east and went into the forest of middling-sized elms. Where it went in I saw a huge heap of dung; I had a good look at it and saw it was made up of vegetable matter.
>
> Some 10 feet up, just where the animal had gone into the forest, I saw a sort of row of broken branches, made, I don't doubt, by the

monster's enormous head as it forced its way into the place where it had decided to go, regardless of what was in its path.

I followed the track for days and days. Sometimes I could see where the anmal had stopped in some grassy clearing and then had gone on forever eastward.

Then, one day, I saw another track, almost exactly the same. It came from the north and crossed the first one. It looked to me from the way they had trampled about all over the place for several hundred yards, as if they had been excited or upset at their meeting. Then the two animals set out marching eastward, one following some twenty yards behind the other,* both tracks mingling and plowing up the earth together.

I followed them for days and days thinking that perhaps I should never see them, and also a bit afraid, for indeed I didn't feel I was big enough to face such beasts alone.

He had a good hunting gun, which would take ball as well as shot, but he had only five cartridges loaded with ball left. All the same he followed the trail as fast as he could, and thought from the freshness of the tracks that he was gaining on the beasts. Meanwhile it was growing bitterly cold, and he had no way of getting warm in the evening except by drinking scalding tea and building a sort of tent of leaves and branches each night.

One afternoon [he went on] it was clear enough from the tracks that the animals weren't far off. The wind was in my face, which was good for approaching them without them knowing I was there. All of a sudden I saw one of the animals quite clearly, and now I must admit I really was afraid. It had stopped among some young saplings. It was a huge elephant with big white tusks, very curved; it was a dark chestnut color as far as I could see. It had fairly long hair on the hindquarters, but it seemed shorter on the front. I must say I had no idea that there were such big elephants. It had huge legs and moved very slowly. I've only seen elephants in pictures, but I must say that even from this distance (we were some 300 yards apart) I could never have believed any beast could be so big. The second beast was around, I saw it only a few times among the trees: it seemed to be the same size.

All this time he had been hiding behind a big larch. When evening came he reluctantly left his point of vantage, because he

* This could not possibly have been deduced from the tracks, and could only be based on what he subsequently saw of the beast in motion. The hunter may well have been carried away by his story; and in any case we should not forget that Gallon, who certainly did not take this tale down verbatim in the middle of a meal, must have retold it and perhaps embroidered it himself.

could no longer bear the cold. Next morning when he returned the beasts had gone. Winter had set in and the weather was too bitter for him to go on tracking them.

> Such [Gallon remarks] was the tale of this man who was too ignorant to know that what he had actually seen were mammoths. And when I told him the name, he did not show the least sign that he understood what I meant.

This report is very different from those of the tribes who live on the shores of the Arctic Ocean where mammoths' carcasses have merely been seen emerging from melting glaciers. The best evidence, to my mind, of the truth of the story, is that the hunter says he met his mammoths in the forest, a place which is not usually supposed to be their natural habitat. Had he—or Gallon—been inventing the story he would surely have tried to give his tale authenticity by putting it in the traditional setting that one sees in all the reconstructions: the great hairy beast advancing with heavy steps through a desert of snow. This is a very unusual background for the animal.

If mammoths happened to leave the taiga in fine weather and go out to pasture on the flowers in the tundra it was often at peril of their lives. For they could easily be taken unawares, especially in September, by the first falls of snow or a sudden frost. Most of the carcasses found in a perfect state of preservation are victims of winter which hid treacherous crevasses and marshes under a white blanket of snow. As the temperature gradually fell, the eternal ice spread, driving the forest and its huge inhabitants southward and imprisoning for thousands of years the straying mammoths that had been killed by the sudden arrival of the last winter. But what of the mammoths that did not stray? Why should they not have survived?

PART SIX

THE TERRORS OF AFRICA

Ati, bwana! There is a story you will not believe, because you are a white man. White men laugh at the stories told by the black men. They say this is not so, and that is not so. We have not seen this or that, so how can it be? They say, Ho, Ho! Black men are like little children, telling tales to each other in the dark. But remember, *bwana,* white men have been in this country for a time that is less than the life of one man, so how can you know all the things that have been known to black men for a hundred lifetimes and more?

ALI, an old African hunter quoted in ROGER COURTNEY'S *A Greenhorn in Africa.*

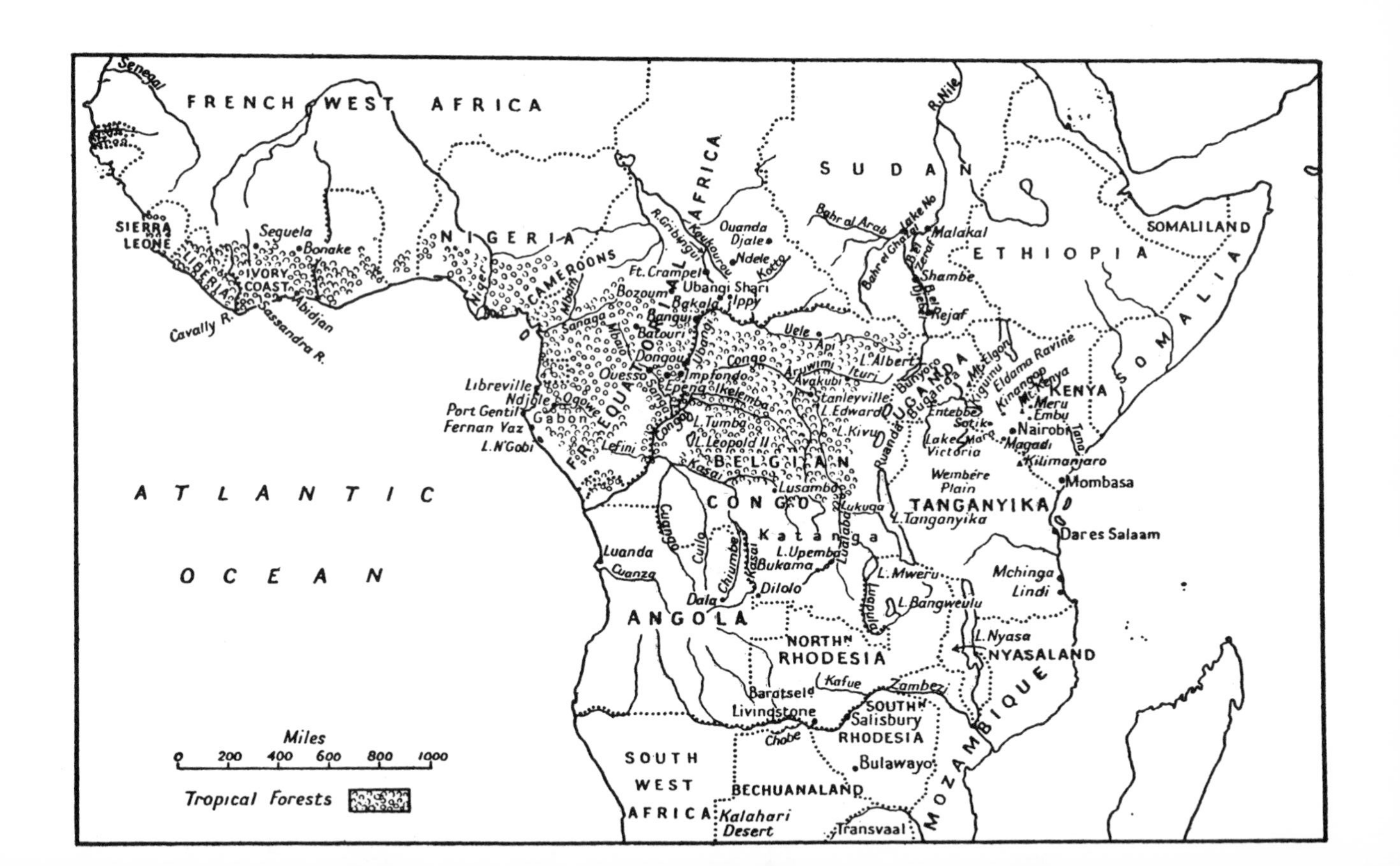
FRENCH WEST AFRICA
Senegal
SIERRA LEONE
LIBERIA
IVORY COAST
Seguela
Bonake
Abidjan
Cavally R.
Sassandra R.
NIGERIA
Niger
CAMEROONS
Mbam
Sanaga
Mbaio
EQUATORIAL AFRICA
R. Gribingui
Koukourou
Ft. Crampel
Bozoum
Ubangi Shari
Bakala
Ippy
Bangui
Batouri
Dongou
Ouesso
Impfondo
Epena
Ubangi
Sanga
Libreville
Ndjole
Ogowe
Port Gentil
Fernan Vaz
Gabon
L. N'Gobi
Lefini
Ouanda Djale
Ndele
Kotto
SUDAN
Bahr al Arab
Bahr el Ghazal
Lake No
Malakal
Zeraf
Shambe
Djebel
Rejaf
R. Nile
ETHIOPIA
SOMALILAND
SOMALIA
KENYA
Mt. Kenya
Meru
Emby
Nairobi
Tana
Magadi
Kilimanjaro
Mombasa
Eldama Ravine
Kinangop
Mt. Elgon
Kigumu
UGANDA
Bunyoro
Buganda
Entebbe
Sotik
Lake Victoria
Mara
Uele
Api
Congo
Aruwimi
Ituri
L. Albert
Avakubi
Stanleyville
L. Edward
L. Kivu
Ruanda
Ikelemba
L. Tumba
L. Leopold II
Kasai
BELGIAN CONGO
Lusambo
Lukuga
Katanga
L. Upemba
Bukama
Lualaba
Dilolo
Chiumbe
Cuilo
Cuango
Dala
Luanda
Cuanza
ANGOLA
TANGANYIKA
Wembere Plain
L. Tanganyika
Dares Salaam
Mchinga
Lindi
L. Mweru
Luapula
L. Bangweulu
L. Nyasa
NYASALAND
MOZAMBIQUE
NORTH^{N} RHODESIA
Kafue
Zambezi
Baratseld
Livingstone
SOUTH^{N} RHODESIA
Salisbury
Chobe
Bulawayo
SOUTH WEST AFRICA
BECHUANALAND
Kalahari Desert
Transvaal
ATLANTIC OCEAN
Miles
0 200 400 600 800 1000
Tropical Forests

CHAPTER 14

THREE LARGE PYGMIES: THE FOREST RHINOCEROS, THE WATER ELEPHANT, AND THE SPOTTED LION

There's a sort of tiny potamus, and a tiny nosserus too—
. . . And I think there's a sort of a something which is called a wallaboo—
A. A. MILNE, *When We Were Very Young.*

"IN AFRICA," Trader Horn has sagely remarked, "the Past has hardly stopped breathing." Several legends of unknown animals seem to refer to creatures of a long-vanished age and are rejected by skeptical scientists on that account. Yet the Negroes have not had lessons in paleontology and they could not know what prehistoric animals looked like. Moreover, animals thought to be extinct are more likely to have survived in Central Africa than elsewhere. Its climate has remained unchanged since time immemorial. Geologically it has not changed for tens of millions of years, and it is the only continent which has not been largely submerged at some time since the beginning of the Primary era.

As Dr. Frechkop has pointed out, the fact that animals like the okapi, forest hog, and water chevrotain are found in Africa proves that "the equatorial forest is the nature reserve of primitive forms which have not been able to survive elsewhere." As it is almost impossible to penetrate this dense, dark, and unhealthy jungle without disturbing the shy creatures that inhabit it, many of them may live there without our having the least chance of meeting them except in exceptional circumstances. Only the natives are likely to know of some of these animals. So it is natural that the legends likely to interest us are mainly found along the vast equatorial rain forest.

When Hans Schomburgk brought a pygmy hippopotamus back alive to Europe in 1913, and thus proved that at least one native legend was well founded, he was probably more inclined to pay heed to others, one of which referred to a sort of pygmy rhinoceros in the mountainous part of Liberia. At first he thought the tales must refer to the forest hog, a big black pig with large tusks, a variety of which is in fact found in eastern Liberia. But he soon found that the Kroo made a clear distinction between the large forest boar and the little mountain rhinoceros, and he eventually came to believe they were right, though he never found a specimen.

There is a good argument to show that this part of Africa may still conceal a pygmy rhinoceros. Professor Trouessart has said of the pygmy hippopotamus:

> The presence of this pygmy species in this particular area in the loop of the Niger leads one to suppose that this country was an island separated from the continent at the time when the Atlantic still penetrated the north of West Africa, that is to say during the Eocene.

It has long been noticed that land animals which live on islands are generally smaller than similar species living on the nearest continent. If the country in the loop of the Niger was once an island, one might expect its ancient fauna to have dwarf characteristics. If there were pygmy hippopotamuses, why not pygmy rhinoceroses?

No *known* rhinoceroses, fossil or living, are found in the area south of the Niger and Senegal rivers. This is odd, since the hippopotamuses and the elephants (which like them are pachyderms of similar habits, dating from the same geological period) spread over this territory. Perhaps the rhinoceroses in this area have remained unknown, because they are so small and can shelter in mountains covered with thick forests. The natives of Liberia are ignorant of this paleontological argument, so it is all the more interesting that they should put the pygmy rhinoceros in the one place in Africa where we have reason to expect to find it.

Liberia is not the only part of Africa where rhinoceroses have been unexpectedly reported in thick forest. Lucien Blancou collected several reports of rhinoceroses in thick forests in the Cameroons and on the Middle Congo River. In such a very peculiar habitat, the rhinoceros could hardly belong to either known genus. Blancou writes:

> The Africans in the north of the Kellé district, especially the Pygmies, know a forest animal larger than a buffalo, almost as large

> as an elephant, but which is not a hippopotamus. Its tracks are only seen at long intervals, but they fear it more than any other dangerous animal. The sketch of its footprint which they drew for M. Millet is that of a rhinoceros. On the other hand they do not seem to have said that it has a horn, though they have certainly not said that it has not. . . .
>
> Around Ouesso the natives talk of a big animal which does have a horn on its nose—though I don't know whether it has one or several. They are just as afraid of it as the Kellé people.
>
> Around Epéna, Impfondo, and Dongou, the presence of a beast which sometimes disembowels elephants is also known, but it does not seem to be so prevalent there now as in the preceding districts. A specimen was supposed to have been killed twenty years ago at Dongou, but on the left bank of the Ubangi and in the Belgian Congo.

Blancou admits that all the evidence is pretty thin; but points out that the evidence upon which the okapi, forest hog, and Congo peacock, and many other creatures were discovered was equally thin. He remarks that in the great wet forest where several of the least-known large species of mammals survive, there seems to be an "ecological niche" for a well-adapted rhinoceros. The red buffalo of the forests is smaller than the black buffalo of the savanna, and probably the African forest rhinoceros is smaller than its brothers on the steppes; perhaps it is a true pygmy like its hypothetical Liberian cousin.

The African elephant is exactly parallel to the African rhinoceros. There are three sizes, one of which is still excluded from the official list. The first (*Loxodonta africana oxyotis*) has rather triangular ears, while in the second (*Loxodonta africana cyclotis*) they are definitely rounded. Both types are spread over most of Africa, the pointed-eared elephant preferring to live in the savanna or bush, while the round-eared one chooses the forest. Moreover the first is much larger than the second. A male bush elephant often stands more than 11 feet high and may occasionally reach as much as 12 feet 8 inches, while the male forest elephant is never more than 10 feet high and is generally about 8 feet.

Besides these two there seems to be a third much more controversial species, the pygmy elephant, in which the adult does not exceed 6 feet 6 inches at the withers. There is no doubt that such elephants exist. Specimens have lived in the Antwerp and New York Zoos, but some zoologists maintain that this does not prove they are separate species; they may merely be the freak offspring of normal elephants. To settle the question one would have to observe the reproduction of pygmy elephants or prove that there are whole herds of them.

64. Forest elephant compared with the larger savanna elephant. Note the different shape of the ears.

The pygmy elephant was first officially described in 1906 by a German professor, Theodore Noack of Brunswick, from a specimen which was six years old and came from Ndjolé in Gabon. He called it *Loxodonta pumilio*. In 1923 Hornaday reported it in the same area, around Lake N'Gobi, and in 1926 Pohle reported it in Fernan Vaz. Not long ago Professor Bourdelle, one of the world's greatest experts on mammals, and Francis Petter published a remarkable study of a pygmy elephant, killed on March 11, 1948, at Aloombé on the coast of Gabon between Libreville and Port-Gentil. It was a solitary old male, thought to be very aged. It was barely 6 feet 5 inches high.

In the Congo the pygmy elephants are often said to be aquatic. A settler living near Lefini showed Hans Schomburgk a piece of thick skin like an ordinary elephant's but covered with a fleece of reddish hairs. He said it came from an animal which the natives called "river elephant" and which had amphibious habits like a hippopotamus. There were persistent rumors all round Lake Leopold II about this curious animal.

Lieutenant Franssen, a Belgian officer, promised to make a systematic search for the animal and send a specimen to the Congo Museum at Tervueren in Belgium. "If the animal exists," he said, "I will come back with it; otherwise I shall not come back at all."

He came back, as he had promised, with the remains of a water elephant captured "after spending thirty-six hours in mud and water." Exhausted by his sufferings on this expedition, and shaken with violent fever, he did not long survive his return. Dr. Henri Schouteden named the water elephant, or *waka-waka* as the natives called it, *Loxodonta fransseni* in his honor. Although his specimen was one of the largest in the herd it was only 5 feet 5 inches at the withers. Its tusks were 2 feet 2 inches long and together weighed under 43 pounds. They were thus considerably longer than those of an ordinary elephant of the same size.

According to Dr. Schouteden the shape of the ears and the two fingerlike appendages on the tip of the trunk (*L. pumilio,* like the Asian elephant, is supposed to have only one) are enough to justify a new species of pygmy elephant. Personally I cannot see much difference in the shape of the ears in photographs of *L. pumilio* and *L. fransseni,* and I find it hard to believe that there are any elephants in Africa with only one finger on the tip of the trunk. It is more likely that *L. fransseni* is merely a synonym for *L. pumilio*—which is hard luck on Lieutenant Franssen.

Glover M. Allen maintained in 1936 that pygmy elephants (*L. pumilio* and *L. fransseni*) are merely forest elephants (*L. cyclotis*) which are not yet fully grown. All the same there is no doubt that adult pygmy elephants exist which are less than 6 feet 6 inches high. The specimen at the New York Zoo, whose growth Dr. Hornaday followed to the end, reached a height of only 6 feet 5 inches and a weight of 1 ton 2½ cwt. An adult of the larger species weighs between 5 and 7 tons. There is plenty of evidence that there exist small groups and even large herds of pygmy elephants, so that Allen's view is hard to maintain today.

Thus there are three species of elephants in Africa, corresponding to three different habitats: the bush elephant, naturally the largest and with the most developed ears, as is usual with animals that live in fairly open country; the forest elephant, smaller, darker, and with rounder ears; and the marsh elephant, even smaller and hairier. There is no gap between the species, which has always made it difficult to tell them apart. But to deny that the pygmy elephant exists, even as a subspecies of *Loxodonta cyclotis,* is absurd.

The reader may feel that the evidence for the pygmy rhinoceros is too vague and that the pygmy elephant is too well established to be unknown. No such objection can be made against the spotted lion.

Colonel R. Meinertzhagen, who was the first to collect remains of the giant forest hog, heard of it several times between 1903 and 1908, but not until 1931 was any more official notice taken of it. Then Captain R. E. Dent, a Kenya game warden, saw four of these large cats crossing the path in front of him, between 10,000 and 11,000 feet up above Meru. They seemed to him smaller than ordinary lions and of a very different type. A few months later his boys, who had set leopard traps on the Aberdare Range, came and told him excitedly that they had caught an unusual animal. It was neither a lion nor a leopard, but a sort of cross between the two. Unfortunately they were not able to bring the specimen back.

A little later Michael Trent shot two small lions with strange fur some 10,000 feet up in the Aberdares. He was not a naturalist and did not think their peculiar appearance was important. Luckily an official in the Game Department happened to see the skins and took them back to Nairobi to Captain A. T. A. Ritchie, then Chief Game Warden, who recognized that there was something unusual about them (see Plate 19, following page 146). They were those of a lion and a lioness, for one of them had a mane, though it consisted of little more than side whiskers.

But Kenneth Gandar Dower, a big-game hunter who also examined the skins, wrote:

> They appear to belong to lions two or three years old—the male had a whiskery mane—and yet the cub spots with which almost every lion is born showed no signs of fading. Certain freak lions do keep their spots to an advanced age, but not in a degree comparable with these rosettes which were distributed not only on the legs and flanks but right up to the spine itself . . .

They so convinced Gandar Dower that there was a separate species of spotted lion, that in 1933 he organized a safari to try to find a specimen of this unknown animal. He found nothing but some footprints too large for a leopard's and too small for a lion's; nor could they be mistaken for a young lion's.

Undaunted, he published an exciting book about the spotted

lion in 1937. As a result G. Hamilton-Snowball wrote to the *Field* about the *marozi,* a spotted animal smaller than a lion, which one usually met in pairs, though this happened rarely and only at high altitudes.

> In the spring of 1923 . . . about 4 p.m. in poor light (it was drizzling a bit) at 11,500 ft. I saw what I thought at first were two very tawny and washed out looking leopards about 200 yards away.
>
> As I turned to my bearer for my rifle . . . I heard an unfamiliar name being excitedly murmured by all the boys who had seen the beasts as soon as I had.
>
> "Marozi, Marozi," they repeated. As I pushed off the safety catch, with one movement, the beasts (or cats) turned and in two bounds had gone into the belt of forest in front of which I had surprised them.
>
> By now I remembered what Marozi meant and asked the boys if they thought it was likely a couple of lion would ever come up from the plains to such a height and in such cold conditions! "Certainly not," they said, "but Marozis live here!"
>
> . . . I must add that the "pugs" certainly were those of lion, and not of leopard, that the beasts looked spotty and tawny, but except for the natives, I never could get confirmation that there even were such things as "spotted lions."

Raymond Hook, the hunter who accompanied Gandar Dower's expedition, thinks that the *marozi* is largely mythical, yet adds that in theory it is not impossible that a small race of lions might survive in a few confined areas, but that there is not enough evidence to prove its existence.

Powys Cobb, on the other hand, is convinced that a species unknown to science lives in the thick and little-explored Mau forest. He surprised a strange beast of the cat tribe, between a leopard and a lion in size, which had attacked cattle grazing in a corner of his farm near the edge of the forest. He mounted his horse and pursued the animal into the forest, where he had to give up the chase without having come any nearer to it. Its footprints were like a small lion's.

The natives of Embu told Major G. St. J. Orde-Brown about "a lion-like forest cheetah" on the southeastern slopes of Mount Kenya. There is no doubt that this is our "spotted lion," for the cheetah is like the lion in outline: both have high shoulders and usually walk with the head raised, while the leopard holds its head farther forward and lower than its shoulders.

E. A. Temple-Perkins, a big-game hunter who is usually inclined to be skeptical about unknown animals, tells of a similar animal.

> The voice I heard that evening at Kichwamba was strangeness itself, and beyond my power to describe with sufficient accuracy, although I heard it at ranges of about one mile to fifty yards from my camp intermittently for two hours. All I can say is that it sounded harsh and guttural; it was not the cough of a leopard, but more like that than anything else if you add the word liquid or gurgling.

Four natives all independently told him that it was the *ntarago*. He was unable to catch sight of the beast in the firelight, nor could he find any spoor.

For some fifty years, then, rumors have been current about this strange cat in Upper Kenya above Meru. Each witness has reported the same features: the size is smaller than an adult lion and nearer that of a cheetah, the spots are like a leopard's, and there is little or no mane.

Most naturalists are apt to be reserved about the spotted lion's existence, but not ill-disposed to it. Reginald Pocock, a great authority on the felines, is distinctly favorable, and brings grist to Gandar Dower's mill:

> There is, as a matter of fact, some independent evidence, unknown to Mr. Gandar Dower at the time, of the existence of a small lion in Kenya. Several years ago, Messrs. Rowland and Ward showed me the skulls of an adult lion and lioness received without skins or more precise locality from that Colony. Their unusually small dimensions puzzled me a good deal. . . .
>
> Since these skulls are decidedly smaller, sex for sex, than any out of the very large numbers that have been measured from many localities in the plains of East Africa, it seems probable that they came from some place in Kenya where few sportsmen have shot and preserved lions. That place may have been the mountain forests of the Aberdares . . .

Colonel C. R. S. Pitman, a game warden and one of the most knowledgeable experts on African fauna, is also favorable.

The skeptics are less convincing. G. Flett of Aberdeen attempts to explain the *marozi* as an optical illusion:

> On two occasions during my two years' continuous stay in the remoter parts of Kenya, I met with lions at very close range during the middle of the day; on both occasions the bodies of the animals were patterned by the shadows thrown by the tree branches and the shadow effect was indeed striking. I thought then that I had seen a species of lion unknown to science and I can well understand how Mr. Pollard's friend was deceived.

But how does he explain the fact that Michael Trent's lions did not change their spots even after they were shot, skinned, and examined by numerous naturalists?

Major W. R. Foran, another skeptic, writes:

> On two different occasions in 1906 I shot a lion and a lioness on the slopes of Mount Kenya, each definitely not younger than five years old; the spotting was clearly visible on belly, flanks and legs. Both were somewhat smaller than the average lion and the male was quite maneless. It struck me that these animals were abnormal in the retention of their spotting, but were possibly a rare occurrence.

What makes him think these were individual freaks and not members of a different race? Elsewhere he makes a remark which contradicts this point of view:

> There is nothing unusual in lions being found at such high altitudes: they are customarily styled "forest" lions, and on average, they are smaller than normal beasts found on the plains at lower altitudes.

Certainly it is not unusual to find lions on mountains, but to find them in thick forests like those on Mount Kenya is quite extraordinary. The lion is an animal of the savanna. It never ventures into thick forests. So if one finds lions in such forests they must therefore be a very special kind of lion. Since racial differences are always shown in differences in form and color, it would be quite normal for forest lions to be smaller and have a spotted coat.

There is thus little doubt that a race of lions with juvenile characteristics could have arisen in Kenya, where it has taken refuge in the thick forests at high altitudes. Cut off there from the lions of the plains, they must have undergone considerable changes as a result of segregation. They seem to have dwindled appreciably, for they are generally said to be between the size of a lion and a leopard. If a large lion (omitting the tail) is 5 feet 3 inches long, and a large leopard is 4 feet, an adult spotted lion would be about 4 feet 8 inches from the tip of the nose to the root of the tail. Perhaps its size is even smaller, since it has several times been compared to a cheetah, which rarely exceeds 3 feet 3 inches, in which case one could clearly call it a "pygmy lion," and it would deserve to be raised at least to the status of a new species. Its exact position in the zoological scale cannot be settled until its skeleton and perhaps also its internal anatomy have been studied. And we have yet to capture a specimen.

CHAPTER **15**

THE NANDI BEAR, AN EAST AFRICAN PROTEUS

Shouting we seize the god: our force t'evade
His various arts he soon resumes in aid:
A lion now, he curls a surgy mane;
Sudden, our hands a spotted pard restrain;
And last, sublime, his stately growth he rears,
A tree, and well-dissembled foliage wears.

ALEXANDER POPE'S translation of HOMER'S description of Proteus.

THE BRAVEST NATIVE HUNTERS IN KENYA will tremble at the name of the *chemosit.* It is said that no one who has heard its hideous cry ever forgets it. But if you ask a settler, whether he be game warden or professional zoologist, he will give you nothing but a frank denial or a skeptical shrug of the shoulders. Yet it would need a whole book to detail all the tales about a sort of large "bear" of unparalleled ferocity found in the Nandi country.

Now the odd thing is that there is virtually not a single species of bear in the whole African continent. Of course there is no certainty that the Nandi bear is really a bear, but it looks like one; and the Wa-Swahili have long used the word *duba,* presumably derived from the Arabic *dubb* (bear), to describe a fearful beast that haunts the forests along the river. Herodotus reports that in Libya there were "huge serpents [pythons] . . . the lions, the elephants, the bears, the asps, and the horned asses. Here too are the 'dog-heads' [baboons]" and many other animals of the local fauna.

It was not until 1905 that Europeans first saw what came to be known as the Nandi bear; and their accounts were not published until there was a particularly persistent crop of rumors in 1912. Geoffrey Williams, who had taken part in the famous Nandi Expedition at the beginning of the century, writes:

I am asked to described the strange beast that I once saw up on the Uasingishu, and it is with some diffidence that I make the attempt since experience has taught me that descriptions of unrecorded animals do not meet with much credence. . . .

Several years ago I was traveling with a cousin on the Uasingishu . . . There was a thick mist, and my cousin and I were walking on ahead of the safari with one boy when, just as we drew near to the slopes of the hill, the mist cleared away suddenly and my cousin called out "What is that?" Looking in the direction to which he pointed I saw a large animal sitting up on its haunches not more than 30 yards away. Its attitude was just that of a bear at the "Zoo" asking for buns, and I should say it must have been nearly 5 feet high. It is extremely hard to estimate height in a case of this kind; but it seemed to both of us that it was very nearly, if not quite, as tall as we were. Before we had time to do anything it dropped forward and shambled away towards the Sirgoit with what my cousin always describes as a sort of sideways canter. The grass had all been burnt off some weeks earlier and so the animal was clearly visible.

I snatched my rifle and took a snapshot at it as it was disappearing among the rocks, and, though I missed it, it stopped and turned its head round to look at us. It is in this position that I see it most clearly in my mind's eye. In size it was, I should say, larger than the bear that lives in the pit at the "Zoo" and it was quite as heavily built. The fore quarters were very thickly furred, as were all four legs, but the hind quarters were comparatively speaking smooth or bare. This distinction was very definite indeed and was the first thing that struck us both. The head was long and pointed and exactly like that of a bear, as indeed was the whole animal. I have not a very clear recollection of the ears beyond the fact that they were small, and the tail, if any, was very small and practically unnoticeable. The colour was dark and left us both with the impression that it was more or less of a brindle, like a wildebeeste, but this may have been the effect of light.

The mist had entirely lifted and the beast stood out quite plainly against the rocks. Unfortunately I had no second cartridge ready, and it had vanished before I could get another shot. Owing to the extreme hardness and dryness of the soil at the time and the number of stones there was no definite spoor to follow, and though we sought for it for some time we never saw it again.

When Williams asked the boy who had been with him to pick out the beast from a book full of illustrations of animals, he plumped for the brown bear. Another Nandi boy confronted with Kipling's *Jungle Book* identified Baloo, the Indian bear, as *chemosit,* a beast whose description Williams found tallied with the unknown animal he had seen.

I have also heard [Williams adds] of the same thing in the Kakumega country near Kabras [about 20 miles south of the Uasin Gishu], and was twice warned by the people not to sleep with my tent door open for fear of the "Shivuverre," which they describe as a nocturnal beast something like a hyaena only infinitely larger and very savage. I heard of a skin in Kabras and tried very hard to obtain it; but could not do so.

The Nandi say that they once killed one years ago owing to its having climbed upon the roof of a hut and broken through.

It killed the people in the hut, but others burnt the place down with the animal inside. They say the reason it is never killed is because it is entirely nocturnal, is very rare, and only attacks solitary people, who never return to tell the tale. . . . the beast was not the least like either a baboon or an ant-bear, of this I am quite certain.

The tales which were current when this very circumstantial account was published are strangely confusing, and the reader may well be puzzled to find me citing reports which unquestionably refer to different animals; for the settlers invented the name Nandi bear and were gradually led into using it for any mysterious—or apparently mysterious—animal they failed to recognize. But these discrepancies are a vital part of the problem. Too many authors have picked out a set of consistent reports, discarding the others, in order to prove that it is some known animal.

A whole set of reports were collected by C. W. Hobley. Major Toulson tells how in 1912:

It was getting dark when one of my boys came into my room and said that a leopard was close to the kitchen. I rushed out at once and saw a strange beast making off: it appeared to have long hair behind and it was rather low in front. I should say it stood about 18 in. to 20 in. at the shoulder; it appeared to be black, with a gait similar to that of a bear—a kind of shuffling walk. Unfortunately it was nearly dark at the time and I did not get a fair view of the head.

Several Dutchmen had asked me a few days before what the strange animal was on the plateau; they said it was like a bear, but they had only seen it at dusk; it turned on their dogs and chased them off. They described it as a thick-set beast and it was making a peculiar moaning cry.

In March 1913 N. E. F. Corbett was fishing just below Toulson's farm, and walked right into the beast:

It was evidently drinking and was just below me, only a yard or so away. I heard something going away and it shambled across the stream into the bush. The place was overgrown and I was without my specs, so could not get a very good view, but am certain that it was a beast I have never seen before. Thick, reddish-brown hair, with a slight streak of white down the hind quarters, rather long from hock to foot, rather bigger than a hyena, with largish ears. I did not see the head properly; it did not seem to be a very heavily built animal.

The next reports come from people building the Magadi railway. They found footprints of a strange animal on the rough track which they had cleared before beginning to lay rails. One of them, called Schindler, who discovered particularly clear prints in the dried mud where a pipe line ran out, drew a sketch which is reproduced here.

An engineer called Hickes who was in charge of building the railway saw the animal itself:

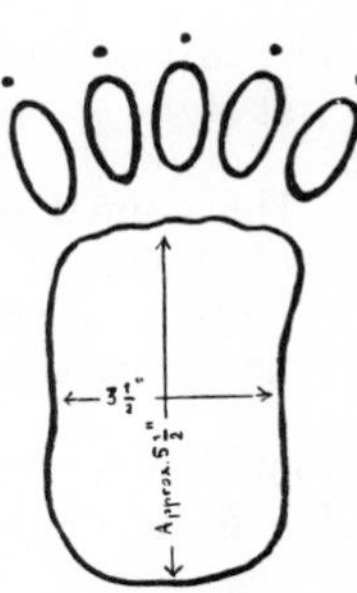

65. Footprint of the "strange beast" found by Schindler on the Magadi railway.

On March 8, 1913, I was travelling alone on our motor trolley along the Magadi railway. At about 9 a.m., when I was at Mile 16, I saw, about fifty yards ahead of me, what I took to be a hyena. . . .

I wondered at seeing a hyena out so late in the morning, and looked at it with interest, especially as, owing to the speed I was travelling (about twenty-five miles per hour) I should pass so close to it before it had time to get away. . . .

As I got closer to the animal I saw it was not a hyena. At first I saw it nearly broadside on: it then looked about as high as a lion. In colour it was tawny—about like a black-maned lion—with very shaggy long hair. It was short and thick-set in the body, with high withers, and had a short neck and stumpy nose.

It did not turn around to look at me, but loped off—running with its fore-legs and with both hind legs rising at the same time.

As I got alongside it, it was about forty or fifty yards away, and I noticed it was very broad across its rump, had very short ears, and had no tail that I could see.

As its hind legs came out of the grass I noticed the legs were very shaggy right down to the feet, and that the feet seemed large and were, of course, covered with black mud. . . .

This strange beast was first mentioned to me by Mr. Clifford Hill, who, on the first survey of this railway, had a young Dutch boy with him who came across one on the Koora Plains (Mile 71).

He had seen nothing like it and could not describe it, so Mr. Hill showed him a picture-book of animals, and he picked out the bear as being like the animal he had seen. . . .

A native servant of one of the engineers, Mr. Archibald, also reported that he saw this strange animal, which, he says, stood on its hind legs and looked at him, but would not run away.

The only other instance of its actually having been seen is reported by a sub-contractor, Mr. Caviggia, who saw one at Mile 38, and his description is very similar to mine.

I have been in Africa—East, South, and West—in the wilds in advance of civilisation, building railways during a considerable part of the last eighteen years, and I cannot think of any animal I have not seen in its wild state, but I have never before seen anything like this beast.

The last of Hobley's reports, about the *koddoelo,* does not seem at first sight to refer to the animal in question.

The animal was described to the District Officer by a Pokomo (who, however, admitted that he himself had not seen it) as being as large as a man, as sometimes going on four legs, sometimes on two, in general appearance like a huge baboon, and very fierce.

Soon Hobley was able to give a more detailed description of the *koddoelo* according to the Wa-Pokomo:

Colour, reddish to yellow; length, about 6 feet; height, about 3 feet 6 inches at the withers; hair long, and all accounts agree on the point of a thick mane; tail short and very broad; claws very long; head, fairly long nose, teeth long but not so long as lion; fore-legs said to be very thick.

The Pokomo state that several have been killed, and one man says that he killed one himself a good many years ago. It is said to be very fierce, and to visit villages and carry off sheep. On these occasions the natives either cross the river until it leaves the neighbourhood or frighten it away by beating drums. The Waboni hunters know the beast well, but say that they prefer to leave it alone. . . .

The Assistant District Commissioner on the Tana also sends a further account of the animal . . . described to him by Pokomo, who said they had seen it . . . Light in colour, long hair on neck and back, usually goes on fore-legs but can go on its hind-legs, not known to climb trees, rather smaller than a lion, tail about 18 inches long and some 4 inches broad, is nocturnal in its habits, fore-legs very thick; said to leave a track with one deep claw mark behind the marks of its four toes (this is rather obscure).* They are agreed about its ferocity and say it attacks a man on sight.

* Perhaps it should be "beyond the marks of each of its four toes," or it may mean that there is a mark of a fifth toe, bent and perhaps opposable like a thumb, possibly more deeply impressed, further back on the foot. In the latter case the beast must indeed be an ape or baboon.

Little by little it began to appear that an animal like a bear in some respect or another was well known to the natives of Kenya since time immemorial. On the coast the Moslem Swahili peoples called the mysterious beast by a name derived from the Arabic for bear: *duba.* In the uplands the Lumbwa name is *geteit* which means "brain-eater," and on the slopes of the Mau escarpment and in the Nandi country it is also called *chemosit* and *chimiset* or "the devil."

The native descriptions are often very exaggerated, or perhaps one should say poeticized. It is spoken of as "great, white and hairy, that walks upright like a man and eats only the brain of its victims," and "half like a man and half like a huge, ape-faced bird, and you may know it at once from its fearful howling roar, and because in the dark of night its mouth glows red like the embers of a log."

There seems to be a confusion in the Nandi mind between a real animal and a devil in their local mythology. Perhaps this is because they call the beast *chemosit* in a figurative sense, just as we call a real animal the Tasmanian "devil." In *The Nandi, their Language and Folklore,* published in 1909, A. C. Hollis drew a fantastic picture, some of whose features are apt to turn up in descriptions of the Nandi bear.

> There is also a devil called *Chemosit,* who is supposed to live on the earth and to prowl round searching to devour people, especially children. He is said to be half man, half bird, to have only one leg but nine buttocks, and his mouth, which is red, is supposed to shine at night like a lamp. He propels himself by means of a stick which resembles a spear and which he uses as a crutch.

Subsequently I. Q. Orchardson, who had lived among the Kipsigis (another name for the Lumbwa) for 16 years, maintained that the *chemosit* should not be considered as entirely mythical.

> Among the Kipsigis one finds no belief that the *Chemosit* is a devil or spirit or god or man. . . . Even those who say it has only one leg say that it has no arms and that its footprint is something between that of an elephant and a rhino and insist very strongly that it is only an animal. A few old men even say that they have seen it at night and give descriptions—very wild ones but all quite unlike man or spirit, for they include fur, whiskers ("wawechik" = whiskers or antennae of animals and insects only). The crutch of Hollis's book is quite unknown here for they say how could an animal use a crutch. A curious detail is that the urine of the *Chemosit* is said to be so evil smelling that no man can stay near it. All the young men

deny having seen it or knowing what it is like, but love to tell children and unsuspecting Europeans fancy tales of it. . . .

Chemosit is a fairly common man's name and it is also used by mothers to make their children obey, e.g. "If you don't eat your food the *Chemosit* will come and eat you."

Because mothers frighten children with bugbears in our country, it does not mean that all bears are imaginary beasts. To the Nandi the *chemosit* seems to be sometimes a bugbear and sometimes a real bear—or at least a real animal rather like one.

And it was not only children who were frightened. "Anyone who has lived in a 'Nandi bear atmosphere,' " writes the game warden C. R. S. Pitman, "cannot doubt the reality of the dread the brute inspires." It is supposed to be the most bloodthirsty monster in Africa, and every year it is blamed for the death of many natives found with their skulls smashed in or the top of the head literally torn off.

Charles T. Stoneham writes:

> Men told me it came down to the villages at night and murdered the inhabitants in their huts. It made its entrance through the roof, killed the occupants, and ate their brains. That was one of the beast's peculiarities; it ate only the brains of its victims. Women gathering firewood in the forest would be missed, and later their bodies would be discovered, always minus the tops of their skulls.

In 1919 a farmer in the Lumbwa country called Cara Buxton told a circumstantial story which may perhaps provide the key to this particular problem:

> A short time ago a "Gadett" [obviously the same word as *geteit*] visited this district. . . .
>
> Its first appearance was on my farm, where the sheep were missing. We finally found all ten, seven dead and three still alive. In no case were the bodies touched, but the brains were torn out. . . .
>
> During the next ten days fifty-seven goats and sheep were destroyed in the same way; of these thirteen were found alive. The Lumbwa were all in a state of great terror, and weird stories were told of the "Gadett": how it walked on its hind legs, pulled babies out of huts, and was even able to kill a man. Finally it was tracked to a ravine and killed by the Lumbwa with their spears. It turned out to be a very large hyaena of the ordinary spotted variety. It had evidently turned brain-eater through some sort of madness.

This anecdote can hardly be said to clear up the whole problem of the Nandi bear. The spotted hyena does not agree with the various descriptions of the unknown animal, nor is it completely proved that the beast the Lumbwa killed was the real culprit.

But the most suspicious thing about the Nandi bear is not the way it attacks, but that it attacks at all. Most so-called savage beasts never attack man unless they are wounded, startled, or provoked. Serious naturalists are apt to be skeptical when an animal is said to be a man-eater. The Nandi bear's reputation in this respect is, however, based on unequivocal facts. A. Blayney Percival, the celebrated game warden, is responsible for what is perhaps the only report of a *chemosit* being killed for its misdeeds: it happened in the Maraquet district not long before 1914.

> It is said that at one time one of these animals was so bad that great preparations were made to kill it, and at last it was killed by a party of men who put a dummy man in the doorway of a hut and sat inside and waited till the animal came and tried to take the dummy; it was then shot with arrows.

The following story is more circumstantial: in 1925 a village had suffered so much from one of these brutes' depredations that the villagers asked the Government of Kenya for help. They said that it had carried off a six-year-old girl one night after having cut a hole through the mud wall of her hut. What is more it had managed to force a way through a thorny hedge 8 feet thick and drag away some calves. Now the surprising thing about this is the way the brute reached his victims. Lions and leopards never try to force a way through the thick and thorny fences because their noses and the pads on the feet are far too sensitive and vulnerable. They jump over the fence. A brute which could force its way through must not only have strong burrowing claws but also an armor of thick hair and hide. Only a creature like a bear could perform such a feat.

Captain William Hichens was sent to investigate these raids, and set to work as if he were after a real beast.

> Around the kopje ran several sandy bush-tracks, which led to water-holes. Cutting down the bush between, Kipet's men joined these paths until they made a sandy track completely encircling the kopje. We brushed this track clean and smooth, so that even a beetle walking over it would leave its trail.
>
> Then I went to bed. I had a small khaki-coloured pi-dog named Mbwambi with me, a mongrel, but a ferocious, plucky, little beast, and I tied him up to the door of my tent.
>
> It was well after midnight when he gave a sharp, alarmed, whiny growl and woke me. But before I could get out of bed the whole tent

rocked; the pole to which Mbwambi was tied flew out and let down the ridge-pole, enveloping me in flapping canvas. At the same moment the most awful howl I have ever heard split the night. The sheer demoniac horror of it froze me still, and not for some seconds did I hear the clatter of poles in Kipet's nganasa, which told of his men, having been aroused, unbarring their hut doors.

I heard my pi-dog yelp just once. There was a crashing of branches in the bush, and then thud, thud, thud of some huge beast making off. But that howl! . . . But never have I heard, nor do I wish to hear again, such a howl as that of the chimiset.

A trail of red spots on the sand showed where my pi-dog had gone. Beside that trail were huge footprints, four times as big as a man's, showing the imprint of three huge clawed toes, with trefoil marks like a lion's pad where the sole of the foot pressed down. But no lion . . . ever boasted such a paw as that of the monster which had made that terrifying spoor.

This "terrifying spoor" is not at all like that found beside the Magadi railway, which was undoubtedly that of a plantigrade animal like a bear or an ape. It seems to be that of a digitigrade animal like a cat, or rather a hyena, since its claws were not retracted. Obviously these must be due to two different animals. Roger Courtney describes some footprints which are more like those of Hichens's *chimiset:*

One day, when hunting in the forest, I was shown by a palpitating game-scout a pair of very peculiar animal footmarks. They were two enormous pug marks, the size of dinner-plates, in a soft patch of ground. They were spade-shaped and turned inward. The claws must have been non-retractile, as I could distinctly see the small cuts where they had dug into the earth. The fact that they turned inward revealed a bear-like character; otherwise I would have said that they had been made by the grandfather of all hyenas.

But, then, a hyena enormous enough to leave footprints as big as those would have himself been a fabulous beast.

On one point everyone who has examined a supposed *chemosit*'s track is agreed: it cannot be a feline animal with retractile claws.

Hichens also describes the depredations of an unknown animal in north Cape Province and the Transvaal in the thirties.

The natives call it the *khodumodumo,* or "gaping-mouthed-bush-monster." . . . this marauder invades the kraals and farms, clambers over six-foot palisades which pole in the cattle-byres and stock-pens and then, seizing a sheep, goat, or calf, leaps back over the fence, to disappear with its quarry. . . . Its footprints are "round, saucer-like spoor, with two-inch toenail marks," a pug which has so far puzzled hunters to identify, since it does not fit the paw of any known wild beast that raids stock.

While this may give an indication of the Nandi bear's area of distribution, it tells us little about its possible identity. Major Braithwaite and C. Kenneth Archer, two settlers in Kenya, are more helpful. They saw in the tall grass and shrub of the savanna what at first they took for a lioness, but its profile had more of a snout. Its head was very large. At the withers it was about 4 feet 6 inches high.

> The back sloped steeply to the hindquarters, and the animal moved with a shambling gait which can best be compared with the shuffle of a bear. The coat was thick and dark brown in colour. . . . Finally the beast broke into a shambling trot and made for a belt of trees near the river, where it was lost.

Captain F. D. Hislop describes a beast which certainly sounds like a bear.

> Just at dusk, we were about half a mile short of the point where the Londiani and Eldama Ravine roads fork. Suddenly, loping along the right-hand side of the road which is bordered by forest, we saw a dark grey creature, about three feet high to the back, round and even podgy, with a small pointed head. It moved on all four legs, but with a slight pronounced power in the hind legs. It went on for about fifty yards until it found an opening into the forest and disappeared.
>
> I am unable to say what it was if not a bear. It was not a hyena or a baboon—of that I am certain. I have never seen any other animal like this. Its whole appearance and action was like a bear, and obviously not afraid of the car, as it moved quite sedately until it found a chance to cut into the forest.

Gunnar Anderssen seems to refer to the same animal. One rainy day he heard his boys shouting, and found that a forest pig had been killed by some very powerful kind of beast.

> I took their description of the beast, without making any suggestions to them:
>
> "Very big, with long black hair and a long tail."
>
> "With long black hair, the tail carried like a dog's, the head not very big, but *baya sana.*"
>
> Very big and powerful it must have been from the way the pig

had been handled. As regards black hair, we found long black hair lost in the battle, soft hair: this was not from the pig, which had coarser hair. The boys could not explain in what way the head was *baya sana,* but they all agreed that it was very bad.

I could not get a clear foot-mark in the grass. What I could see looked very large, something like the mark of an old leopard which could not draw in his claws properly. The pig appeared to have been killed in an extraordinary way, as if it had been hit with, say, a log, breaking the backbone; it had then been turned over and the stomach, heart, etc., had been eaten. . . .

This story may sound a bit funny—I think so myself—but so it happened.

Finally, to complete the record, here is a strange encounter which Charles T. Stoneham had at Sotik on the edge of the Lumbwa Reserve.

One night, thinking I heard a noise, I sneaked quickly from my hut and stood against the wall of it, shaded by the thatched eaves. . . . I saw some beast travelling towards me through the mist, and I could not think what creature it could be. . . . Slowly and quietly it approached, and I waited, intensely interested, knowing it could neither smell nor see me. The moon passed behind a cloud; the animal became indistinct in the mist. When within a dozen paces it stopped, and at that moment the moon shone out again clear and brilliant.

I received a dreadful shock. The beast was like nothing I had ever seen or imagined. It had a huge square head, and the snout of a pig; its eyes, two black spots, were fixed upon me in an observant stare. Large circular ears, the size of plates, stood up from its head, and they were transparent—I could see the grass through them. The creature's body was covered with coarse brown hair, its tail was the size of a tree trunk.

There is an ant-eater in Kenya, a survival of the age when the cave bear and the woolly rhinoceros roamed the earth, but this beast, though like to that rare species, was not of it.

When he told his friends about his adventure the next day they all were inclined to think he had seen the Nandi bear.

But I am inclined to think I saw some weird, hybrid ant-eater, and am sorry I had not the pluck to go after it and find out all about it.

That is the chief evidence for the Nandi bear. Now for the experts' opinions. Blayney Percival writes:

The stories vary to a very large extent, but the following points seem to agree. The animal is of fairly large size, it stands on its hind legs at times, is nocturnal, very fierce, kills man or animals.

C. R. S. Pitman, who does not believe that the Nandi bear is an unknown species, has underlined the contradictions in the numerous reports he had collected personally:

> By day, a dark rufous or tawny-hued creature resembling either bear or gigantic ape, almost invariably viewed indistinctly at a distance and usually in longish grass; and by night a long-haired brute having the appearance of a bear, abnormal hyena or South American giant anteater.

And indeed the beast would have to be Proteus himself in order to look like all the animals it has been said to resemble: a gorilla, an anteater, a hyena, and a bear, and to be black, tawny, red, and gray in color, sometimes tailless and sometimes with a brush, sometimes earless and sometimes with distinct ears.

The most obvious conclusion is that it is an animal belonging to one of the types just mentioned, and that all these puzzling divergences are due to people observing other animals which have nothing to do with the question. Let us begin with C. R. S. Pitman's list and proceed by elimination. We can discard the giant anteater at once, for it is a specifically South American animal, belonging to a group found only in America. I feel sure that the animal

66. The aardvark.

actually referred to is the aardvark, whose Latin name is *Orycteropus* and whose English one is the Afrikaans for "earth pig." This animal is very rarely seen, since it spends the day in burrows under the earth and prowls only at night. It has a pig's snout and huge ears. It has a powerful tail, upon which it can lean like a kangaroo, and often stands on its hind legs. Could the Nandi bear be an aardvark?

Certainly it is so rare to see one that people might easily be bewildered by meeting an aardvark. Personally I am almost certain that it was an ordinary aardvark that Charles Stoneham saw, whatever he may think. No doubt he had seen this beast only in drawings and photographs, and in the moonlight and mist it seemed to him even more extraordinary than it really is. His description is so full of features that betray the animal's identity—the pig's snout, round ears that sometimes hang down on the side of the head and have skin so thin as to be translucent, the brown fur, the tail as long as a tree trunk—that there can be hardly any doubt. Stoneham's beast could easily be an aardvark, but none of the other manifestations of the Nandi bear are in the least like it.

It might seem self-evident that the Nandi bear is a bear. Most witnesses compare the *chemosit* to one, and the Wa-Swahili give it the Arab name of *duba.* But, in fact, most of the accounts do not agree with a bear. Many of them mention a clearly visible tail, and the others never speak of a stump of a tail, so characteristic of the bear, but say that it was indistinct or not visible, implying that it could well have been tucked between the legs. There are also descriptions that insist that it has a mane and smoother, less hairy, hindquarters. All the bears are remarkably evenly covered with fur. Moreover, when in a hurry, but not actually galloping, they adopt a gait which is so characteristic that it is not easily overlooked. Now although some witnesses have said that it had "a gait similar to that of a bear—a kind of shuffling walk" or "a shambling gait which can best be compared with the shuffle of a bear," no one has actually said that it *ambled;* that is to say, that it ran by simultaneously moving both legs on one side of the body. This is most important, since no ape runs in this way, and the only carnivores to do so are the bear, the hyena, and the South American maned wolf. That a heavily built animal should shuffle or shamble is quite natural, but so would a gorilla, a bear, an old lion, or a wild pig. The only witness who has described the beast's gait in detail is Hickes, who says it "loped off—running with its fore-legs and with both hind legs rising at the same time." This irregular gait in which the forelegs touch the ground in turn, while the hindquarters move as a single unit, is like the so-called "pithecoid" gallop so characteristic of monkeys in a hurry. Williams's cousin's phrase "a sort of sideways canter" is a good description.

This brings us to the next theory: is the Nandi bear an ape? The Negroes of the Mau and Nandi country usually think of the *chemosit* or *kerit* as a sort of giant ape, and the Wa-Pokomo describe the *koddoelo* as an enormous baboon. The beast that the Nandi villagers killed because it had climbed on the roof of a hut could hardly have been a lion or a hyena. It sounds more like an ape. Even the white men's descriptions often suggest an ape. "In most stories," Blayney Percival points out, "the resemblance to a monkey of sorts is very noticeable, but the fact that the animal is nocturnal, a point on which all native accounts agree, at once makes this impossible."

This a priori argument is not really tenable. Primates as a whole show a clear tendency toward a nocturnal life. The tarsiers and most lemurs have chosen this life and are partly adapted for it. Among the true monkeys, there are several species with nocturnal habits, especially in America.

It would not be so extraordinary if an African ape were to have nocturnal habits. This may even be the reason why it is unknown to science.

Hichens writes about this theory:

> Some of us who have hunted the brute share the view that it may be an anthropoid. Its raids invariably occur on the skirts of forest land, which might be the haunt of one of the great apes. To those who would object that the apes are not man-killing carnivores, the answer is that one is not so sure; the chacma baboon is a desperate "carnivore" and is a serious menace to sheep-farmers in South Africa, where the baboon-packs raid the flocks, ripping up lambs with their long-clawed thumbs and lion-like fangs and carrying off the carcases to their kopje haunts.

Colonel Pitman writes of baboons:

> One year, in a part of the Bunyoro District, five native children were attacked and severely bitten within a few days of each other: two of them died. . . . Children, while protecting crops, are sometimes killed. The ape's method of attack is to seize the little watch-

man in its arms and then virtually disembowel him with terrible, downward strokes of its muscular feet.

E. G. Boulenger adds:

So utterly without regard are baboons for native women that in some remote parts of Natal washing clothes or filling water-vessels can be done only under cover of an armed guard.

I do not think that baboons attack women from mere lack of "regard," but from sheer lust. Certainly apes are sexually stimulated by the sight of women. I have several times seen a young tame baboon, whose master had brought it to the Mediterranean coast, assault women who were sun-bathing and attempt to abuse them. And it was hardly a year old and had not reached the age of puberty! When this happens in nature, and the monkey is an adult male almost as big as man and almost mad when in rut, there could obviously be a bloody scene if the victim tried to resist. A large baboon's teeth are quite as formidable as a hyena's or a leopard's. It is hardly surprising that even the larger beasts of prey avoid these quarrelsome and brutal apes and that natives refuse to venture in the forests where they dwell. Does the so-called "Nandi bear" agree with the description of any kind of baboon? The *koddoelo* certainly does, since the Wa-Pokomo themselves compare it to one. The long hair on the back and neck where it forms a mane, thicker forelegs, long snout, long canine teeth, four powerful claws, short and thick tail: the picture is complete. Even the baboon's gait is described exactly: it can occasionally stand on its hind legs but it very seldom climbs trees.

What about the beast of the uplands, seen on the Uasin Gishu plateau and at Magadi, and called *chemosit* or *kerit* by the natives? If we compare the three best European accounts, Williams's, Hickes's, and that of Major Braithwaite and Kenneth Archer, we find they agree perfectly in the following particulars: *

1. Thick stocky body.
2. High withers, sloping back.
3. Forequarters covered with thick fur, hindquarters smoother and barer.

* I am shelving for the moment reports like Major Toulson's, Captain Hislop's, and that of Anderssen's boys, which all refer to a clearly different animal, all black and with short legs, to say nothing of Corbett's beast which is different again.

4. Long rather pointed snout.
5. Small ears.
6. No visible tail (the mandrill and the drill have tails no longer than a bear's).
7. Color tawny (like a lion or hyena) to dark brown.

Hickes adds that it was broad across the rump, had a short neck, and "stumpy" nose, and that the legs were shaggy right down to the feet which seemed large.

One could hardly wish for a better "pen portrait" of one of the Cynocephala, the dog-headed monkeys which include the gelada, drill, and mandrill as well as the baboons proper. One wonders why Williams insisted that it was not the least like a baboon. Presumably he had the ordinary green baboon in mind, and perhaps the beast's size was the reason, for he estimated that it was almost 5 feet high when sitting up. The largest known baboon, the gelada (*Theropithecus gelada*), is nowhere near this size, being no more than 3 feet high whether sitting up or on all fours.

Now, while there are no known baboons as large as the "Nandi bear" they existed in the past. In India there was once a sort of giant baboon, the *Simopithecus,* which was twice as high as the present baboons. Moreover Dr. Broom has found remains of giant baboons in Africa itself; and he has given them the scientific name of *Dinopithecus*. May not their descendants have given rise to the widespread legends about the *chemosit* and *koddoelo?* At all events a reconstruction of a giant baboon is extraordinarily like most of the natives' and many of the settlers' descriptions of them. And if such a beast survived it is easy to see why the *chemosit* arouses such terror. The mere thought that there may be a living animal as huge and strong as a gorilla and as brutally savage as a baboon is frightening enough.

This theory, which I do not think anyone has put forward before, is too daring for the skeptical zoologists who prefer the more commonplace explanation of a giant hyena, a familiar animal in Africa. This was the view of R. I. Pocock, Superintendent of the London Zoo. He has maintained that the Nandi bear is merely the common spotted hyena (*Crocuta crocuta*), which has a head and general outline strikingly like a bear's and which moreover ambles like a bear. It is also savage enough to be the monster.

The spotted hyena is very cowardly, and often shrinks from attacking herds for fear of their hoofs. But when the hyena is

67. Reconstruction of a giant baboon.

starving it will creep into the native huts to seize someone asleep. The children are usually put to sleep in the passage at the entrance to the hut, and thus become the hyena's prey. Natives often bear the scars of horrible mutilation on their faces, a sign that they were attacked as children. Moreover, the spotted hyena seems to grow fiercer with age:

> Some of the older animals not infrequently become abnormally large [writes Colonel Pitman]. Development in size seems to be coincident with increased ferocity. Tales of "man-killing" usually refer to "giant" hyenas. For instance, a giant hyena dragged an old woman from a hut in Karamoja, and devoured her. In the same district another giant marauder killed and ate a seven-year-old boy, who was sleeping out in the thorn scrub.

The Nandi bear's fur is sometimes described as black, sometimes as reddish. Pocock has pointed out that the spotted hyena varies greatly in color in the course of its growth, and that there are red mutations (erythrism) of the spotted hyena, which may be the origin of the Nandi bear. These are commoner in Somaliland and Sudan than in East Africa, where they might be taken for unknown monsters. Pocock therefore thinks that the red hyena is "the Chemosit, Kateit, and Gereit," and so on. He backs up his argument with a skin sent him from Nyasaland by a man who thought it came from a Nandi bear. Actually it belonged to an unusually red spotted hyena, and the skull sent with it was that of an unusually large leopard.

I do not think this skin clinches Pocock's argument. If he had received only the skin one could conclude that some people at least thought that the red hyena was an extraordinary creature worthy of the name "Nandi bear." But as it is, all we can say is that Pocock's correspondent was either the victim or the author of a hoax.

There have been similar hoaxes about most unknown animals. They are more often the work of some wag trying to confuse the issue than of a forger trying to make money. Sometimes they even come from a convinced believer, who, for lack of proof, fabricates a false one, thus discrediting the well-founded legend he was trying to prove.

Charles Stoneham tells a story which is much better evidence for the hyena theory:

> In 1930 a man brought me a skull like a hyena's skull, but of larger size. He had a farm at Koru, where at about that time cattle and sheep had been killed by some mysterious agency. The Lumbwa herdsmen ascribed these deaths to the chemoiset. The farmer had set numerous traps, and one day, on returning from a week's holiday, his boys informed him that the marauder had been caught. It was a young chemoiset, they said. Being unwilling to lay hands on the beast's body they had tied a rope to it and towed it away into the bush. There the farmer found all that remained of the body, and he brought the skull to me.

This skull turned out to belong to a brown hyena (*Hyaena brunnea*), a smaller beast than the spotted hyena (*Crocuta crocuta*). It is heavily enough built to remind one of a bear. But its plume of a tail and its particularly long and upstanding ears make it difficult for one to see how there could be any confusion.

The brown hyena is not usually supposed to be found north of the Zambezi, but it is not impossible that there may still be very rare specimens in Tanganyika and even in Kenya. Naturally the unusual appearance of one of these dark-maned hyenas would be most disconcerting and liable to produce plenty of fantastic interpretations. And Stoneham's story shows that even the natives could mistake a brown hyena for a young *chemosit*.

On the other hand, Hichens writes of the spotted hyena:

> Hyenas are as common around most African villages as rabbits around English villages; their habits, appearance and spoor are thoroughly well known to every villager and to suggest that, not once, but dozens of times in many places, kraalsmen would mistake a hyena for a Nandi bear is on par with suggesting that countryfolk all over England could mistake a rabbit for, say, a fox.

It is quite clear from almost all the descriptions that the beast cannot be a hyena, unless it is some strange crossbreed with a spotted hyena's round ears and an even shorter tail, combined with the brown hyena's thick dark fur. The hyena is a digitigrade animal and would find it as hard to stand up on its hind legs as a dog or a cat. And a hyena that stood 4 feet 6 inches at the withers would be an unknown animal as remarkable as a giant baboon.

Moreover, as Captain Hichens has pointed out, the *khodumodumo*'s method of stealing cattle in the Transvaal would be impossible for a hyena, which drags its victims. It is as incapable of leaping over a 6-foot fence with a calf in its mouth as it is of forcing its way through a wall of thorns.

So the Nandi bear is not an aardvark, bear, known ape, or ordinary hyena. We have been multiplying possibilities without simplifying the problem. It is the very quantity of evidence that makes it so complex. Indeed, the trouble is that many of the descriptions apply to an unknown animal, and that other known animals have crept in to blur the picture. Let us see what these known animals are.

Here I am glad to be able to offer a very pertinent description of what at first seems to be an unknown animal in the Belgian Congo, given me by Georges Sandrart.

> In 1936 I was on an excursion on the ridge of the Congo–Nile watershed, at an altitude varying between 8,000 and 11,500 feet, where some of the last patches of virgin forest still survive. One day in broad daylight about noon I saw a little bear, slightly smaller than a Carpathian bear, round a bend in the path. I was astounded, so I

took great care to examine it minutely. Its fur was shining blackish brown and I could see a triangle of silver hairs on the peak of its forehead. The apparent absence of tail, the shape of the head, the large snout, the little round ears, the slope of the back, the relatively long legs, *everything about the animal reminded me of a bear*. But a bear in Africa. . . .

He wasn't dreaming. It turned out to be an animal that is rarely met, because it is both nocturnal and burrowing, but which is well known to zoologists: the honey badger or ratel (*Mellivora ratel*), which G. S. Cansdale says is probably seen less often than any other small carnivore. All the same it is found all over Africa as well as in Arabia and India.

The ratel is not a bear, even though it looks like one, and also tends to stand up on its hind legs. It belongs to the closely related family of Mustelidae and is a sort of plump badger, all black except for the top of the head and the back, which are covered with a clearly defined band of silver hair.

Was this then what so many people took for a bear? I must admit that for a long time I believed that the ratel was a small creature, not so big as a European badger, but Sandrart's account set me researching, and I was astonished to find that some ratel skins from Central Africa are as much as 3 feet 6 inches from the

68. When the ratel turns black with age it is very much like a bear.

nose to the tip of the tail (which is quite small and no more than 8 inches long). It is almost as big as the smallest of the bears, the Malayan sun bear.

But if some of the Nandi bears were ratels, why had nobody mentioned the conspicuous band of white? This puzzled me until I discovered that in 1906 Lydekker had described a West African ratel which had no silver band and was therefore entirely black. Therefore in Africa one may meet a little-known animal, looking very like a small black bear; and this explains some of the descriptions of the Nandi bear. There is no doubt that what Major Toulson, Captain Hislop, and Anderssen's boys saw was a black ratel. A "round and even podgy" beast, 3 feet long and between 18 and 20 inches high, with short legs and covered with long black hair, a small pointed head, the tail carried like a dog's with nonretractile claws—what else could it be?

Blayney Percival tells how he had to go and inquire in the Sotik country about an animal very like the Nandi bear. It was supposed to be fairly large—the size of a pointer—and it stood on its hind legs and was very fierce. It turned out when he showed them a picture of the ratel that this was what it was. I think the ratel may also be the answer to the problem of the *too*, an animal Schomburgk heard rumors about in East Africa, "as big as a goat, with teeth like a dog's, dark black fur and an evil character."

Pursuing the question of black ratels I found that the truth was more relevant than I had expected. From the very detailed account which Dr. Welch has given of the changes in color which an old ratel underwent during 12 years in the London Zoo, it seems to be quite normal for ratels to become black at the end of their lives, when their size is largest. The largest ratels are therefore also the blackest, and would naturally not be recognized as the same animal as the young ratels, which are smaller and strikingly marked.

My friend Charles Dewisme has suggested that the legend of the Nandi bear may be based on two separate animals: the spotted hyena and the ratel, the hyena's bloody deeds being attributed to the mysterious little "bear" which is so rarely seen. It is natural for any little-known animal to be blamed for every crime, just as a stranger in a village is always the first to be suspected.

Patrick Bowen has suggested to Frank Lane that the Nandi bear may be a scapegoat in a more sinister sense and that native witch doctors may well be responsible for the massacres attributed to the mysterious animal. They are very clever at imitating animals' cries and making false tracks as a means of stealing cattle and revenging themselves on their enemies with complete impunity. Just as in the Belgian Congo and Nigeria there is a secret society of leopard-men called Aniotos, and in Europe in the Middle Ages there were men who were werewolves, there may well be a sect of bear-men, based on Arab folklore, hyena-men, or even ratel-men in Kenya and Uganda. The Aniotos cover their heads and backs with leopardskins and use mummified paws on the ends of sticks to imitate the beast's tracks. They attack their enemies at night and maul their necks with gauntlets armed with iron claws, thus producing horrible wounds indistinguishable from those made by a large cat. In the same way they might easily imitate the bear's method of killing.

This would explain those native descriptions of the *chemosit* as half man, half bird, with an ape's head and walking upright. And the footprints like those of an old leopard which could not draw in its claws properly could well have been made by a lion's foot stuffed wth straw or sand, or worn on the real criminal's foot. All this is pure conjecture. But the terrible atrocities of the Mau-Mau in Kenya have shown the cruel lengths that the Kikuyu would go to, and make this theory all too possible.

Actually there is no need to consider the ratel as a harmless scapegoat blamed for hyenas' or witch doctors' misdeeds. It is not always so harmless.

Guy Babault, a French naturalist who has been observing the fauna of Central and East Africa for 30 years, writes:

> The courage of this little animal is unparalleled. A lion will let a determined man pass, but not the ratel. When disturbed it will attack and only retreat in order to return more bitterly to the charge. It is wonderfully tenacious and brave. Moreover it is armed with strong jaws and powerful claws, for it is a burrower that can dig itself deep earths.

The ratel belongs to a very fierce family of carnivores. The Mustelidae have not the panther's or the tiger's reputations only because they are smaller. But relative to the size of their victims, they are really much more cruel, fearless, and aggressive than the felines. The wolverine or glutton (*Gulo luscus*) of the arctic, which is no bigger than a large ratel, well deserves its name, since it lives upon large herbivores like the reindeer and even small elk. A large ratel about the size of a wolverine might well kill pigs, sheep, and even cattle; and it might possibly kill a man.

Moreover the ratel is one of the few mammals one might suspect of penetrating a wall of thorns to reach its prey. Its dense fur and a thick layer of subcutaneous fat make it impervious to the stings of wasps and bees when it raids their nests.

In its method of killing its prey by a blow with its paw on the top of the head and then disemboweling, the Nandi bear is much more like an animal with powerful claws, like the ratel, than one with powerful fangs, like the hyena or lion. The ratel is therefore perfectly capable of committing many of the misdeeds attributed to the Nandi bear.

Bears, which are usually peaceable and gentle and live on fruit and insects, sometimes become complete carnivores when they are old solitary males, and none but old bears are ever dangerous to man. Might not the same be true of ratels, whose habits are relatively little known, but which are closely related and very similar in appearance and behavior?

This theory is now a simple demonstrable theorem:

The Nandi bear is described by the Kenya settlers as a small black bear of bloodthirsty habits. But a dangerous black bear is an altogether unusual sight in Africa.

When a ratel grows old it changes so much in its appearance and its nature that it is also a very unusual sight.

Now when an old ratel turns black all over, reaches its full size, and becomes a more bloodthirsty carnivore, it looks and behaves just like a small black bear.

Therefore the Nandi bear is an old ratel. Q.E.D.

Of course this theorem is valid only if you begin by accepting that the Nandi bear looks like a small black bear. As I have already shown, the settlers have often used the name for an animal which seems more like a giant baboon. Moreover the *chemosit* or *kerit* is described by the natives as a big hairy biped, and the *koddoelo* as an enormous baboon. I believe the whole confused question of the Nandi bear was based on a misunderstanding.

When Geoffrey Williams and his cousin saw a strange animal, which I think may have been a giant baboon, they compared it to a bear. When Hickes saw a similar animal beside the Magadi railway, the previous encounter came to mind, and word got around that there was a sort of bear in Kenya. At once a whole string of people, Major Toulson, the Dutchmen on the Uasin Gishu, and

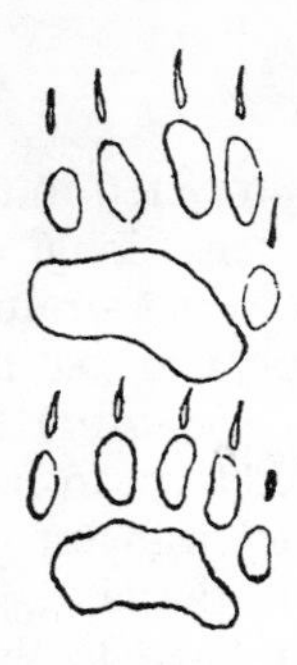

69. Badger's footprints. *Left:* separate prints of forefoot followed by hind foot. *Right:* composite print formed when they overlap.

several other witnesses who doubtless had met large black ratels, confirmed the truth of this rumor, which was supported by the fact that the Wa-Swahili knew an animal which they called *duba* from the Arab name for bear. No doubt this was also the ratel.

Finally some plantigrade footprints were found which might have been made by a giant baboon, by an aardvark, or by a giant ratel, a beast which is almost as plantigrade as a bear. In any case when the ratel puts its hind foot down very close behind its forefoot, as it generally does, it makes a composite print remarkably similar in shape to the one discovered near the Magadi railway.

So the picture of the Nandi bear became blurred and confused, and people began to imagine that any beast they did not recognize must be the mysterious animal. Corbett, handicapped by having forgotten his spectacles, was one of these. He describes an animal not in the least like either a baboon or a ratel, but which may well have been a Cape hunting dog (*Lycaon pictus*) or even a brown hyena. The white spots on this hyena's limbs might account for Corbett's "slight streak of white down the hindquarters." And, of course, the settlers who saw an unusually red hyena, or a little-known brown hyena, added it to the Nandi bear's confused dossier. Any bloody raid made by a hyena or other man-eater, and any slaughter of cattle or men by a "brain-eating" maniac, whoever he may be, would be attributed to the legendary bear without the least proof. And perhaps witch doctors took advantage of this atmosphere of terror to carry out their own crimes, leaving false clues and further confusing the affair.

The Nandi bear is now a bear no more. It is a whole apocalypse of beasts, and an apocalypse as obscure as St. John's. All the same I trust I have been able to disentangle the four horsemen: the giant baboon, the old black ratel, the unfamiliar hyena, the witch doctor.

One other beast may possibly also have contributed to the Nandi bear. In 1923 Dr. E. G. Wayland made a sensational discovery on the Bunyoro side of Lake Albert in a very recent stratum, probably dating from the Pleistocene. This happened on the scene where the *engargiya* (the local version of the *chemosit*) had committed its crimes. The bone that he disinterred was an incomplete phalanx belonging to a member of one of the strangest groups of mammals the earth has ever known, the Chalicotheria.

These Perissodactyls, which are distantly related to the horses, were a headache for the paleontologists for a long time, because, while they have an ungulate's skull and teeth, their feet have three

70. Reconstruction of the Chalicotherium by Professor Othenio Abel, 1920.

toes armed with formidable claws, with which it is now thought they tore down the leaves which were no doubt their food.

The remarkable thing about Dr. Wayland's find was that it was the first time that remains of one of the Chalicotheria had been dug up in Africa. Until then several species had been discovered in Europe, Asia, and even in America, but never in Africa.

When the precious phalanx reached the British Museum of Natural History, it was identified by Dr. Charles William Andrews, who was struck by the apparently recent date of the fossilization. He also recalled that remains of Chalicotheria had been found on the island of Samos, where they were associated with those of the Samotherium. They had shared the same habitat as this Greek okapi, so might they not also have a living representative hidden in the African forests? With their short hind legs and very sloping back, their heavy head and short tail, they must have been very like a hyena in outline—but a hyena as big as a horse. Dr. Andrews therefore wondered whether some belated Chalicotherium might not be the origin of the Nandi bear.

Of course this indisputable vegetarian could not be saddled with the crimes attributed to the *chemosit*—which could often be due to spotted hyenas. But there is no doubt that if a Chalicotherium were clad in a furry coat—which is pure supposition—it would agree perfectly with most witnesses' descriptions.

Science is a long way from officially adding the Nandi bear to the list of African fauna. We have only the most dubious remains of it. Ivan T. Sanderson says that in the British Museum of Natural History there is a piece of fur covered with long brown hair which may come from the animal about which armchair experts are so skeptical, but whose existence it is rash to deny in many villages in Kenya.

CHAPTER **16**

THE LITTLE HAIRY MEN

Up the airy mountain
Down the rushy glen
We daren't go a-hunting
For fear of little men.
WILLIAM ALLINGHAM.

BY FAR the most mysterious of African tales are those about the *agogwe*, little hairy men, barely 4 feet high, which walk on their hind legs. Captain Hichens tells how:

> Some years ago I was sent on an official lion-hunt to this area [Ussure and Simbiti forests on the western side of the Wembare plains] and, while waiting in a forest glade for a man-eater, I saw two small, brown, furry creatures come from the dense forest on one side of the glade and disappear into the thickets on the other. They were like little men, about four feet high, walking upright, but clad in russet hair. The native hunter with me gazed in mingled fear and amazement. They were, he said, *agogwe,* the little furry men whom one does not see once in a lifetime.

S. V. Cook tells of legends of little red men in the hills of Kwa Ngombe, 350 miles further northeast.

> Old Salim, the interpreter at Embu, tells me with great dramatic effect how he and some natives once climbed to near the top when suddenly an icy cold wind blew and they were pelted with showers of small stones by some unseen adversaries. Happening to look up in a pause in their hasty retreat, he assures me that he saw scores of little red men hurling pebbles and waving defiance from the craggy heights. To this day even the most intrepid honey hunters will not venture into the hills.

Cook naturally disbelieved this story and thought it no more than a charming fairy tale. But it supports Hichens's story. So does Cuthbert Burgoyne:

> In 1927 I was with my wife coasting Portuguese East Africa in a Japanese cargo boat. We were sufficiently near to land to see objects clearly with a glass of twelve magnifications. There was a sloping beach with light bush above upon which several dozen baboons were hunting for and picking up shell fish or crabs, to judge by their movements. Two pure white baboons were amongst them. These are very rare but I had heard of them previously. As we watched, two little brown men walked together out of the bush and down amongst the baboons.
>
> They were certainly not any known monkey and yet they must have been akin or they would have disturbed the baboons. They were too far away to see in detail, but these small human-like animals were probably between four and five feet tall, quite upright and graceful in figure. At the time I was thrilled as they were quite evidently no beast of which I had heard or read.
>
> Later a friend and big game hunter told me he was in Portuguese East Africa with his wife and three hunters, and saw a mother, father and child, of apparently a similar animal species, walk across the further side of a bush clearing. The natives loudly forbade him to shoot.

Now there is no known ape which normally walks upright on its hind legs. While a monkey may stand up on its hind legs for a moment, the position is unusual. In fact only the gibbon, which is found only in Asia, is a biped, its long arms acting as balancing rods; but it is specially adapted to swinging from its arms in trees and rarely descends to the ground.

Perhaps the *agogwe* are therefore really little men. Their description seems at first sight to apply pretty well to the African Pygmies. Pure-blooded Pygmies are never more than 4 feet high, and their bodies are not so smooth as those of the other Africans. It would not be so very extraordinary to find them in Tanganyika or Mozambique, where there is thick forest, their favorite habitat. Driven southward and eastward by the Masai, they could easily have reached the great lakes.

Georges Sandrart has suggested to me that the little men could equally well be full-blooded Bushmen, who are hardly taller than the Pygmies, varying in height between 4 feet 8 inches and 5 feet 2 inches. But their very bare and quite hairless skin makes them poor originals for the *agogwe*—hairy little men. In any case the

East African Negro traditions make a clear distinction between the Pygmies and these small bipeds which they think are nearer monkeys. Roger Courtney's guide, Ali, told his master how his own father, who was driving his sheep to pasture on the slopes of Mount Longenot, fell into the hands of these gnomes when he went into a cave. He was stunned from behind, and when he came to he found he was surrounded by strange little creatures. "The Mau men were lower even," he told his son, "than those little people of the forests [the Pygmies] for, though they had no tails that I could see, they were as the monkeys that swing in the forest trees. Their skins were white, with the whiteness of the belly of a lizard, and their faces and bodies were covered with long, black hair." To his great surprise the shepherd noticed that his spear was still lying at his side. "The Mau men who are so nearly monkeys did not know what was the spear. It is possible they did not know I could have fought with it and killed many of them."

The Mau men with their white skin and black hair hardly agree with the brown or reddish *agogwe*, but it has been found time and again that African legends contain a grain of truth. In the Ivory Coast, almost at the opposite end of Africa, there is a legend of reddish hairy dwarfs identical in every respect with that in Tanganyika, as I learn from Professor A. Ledoux, who in 1947 was the head of the Zoological Department of the Institute at Adiopodoumé, 12 miles from Abidjan.

One evening a young African who worked in his laboratory came and told him that one of his colleagues had seen a Pygmy on the previous day.

The professor raised his eyebrows.

"A Pygmy, here?"

"Yes, about 500 yards from here."

The professor nearly fell out of his chair in surprise. "Why didn't your friend come and tell me at once?" he asked suspiciously.

The young man explained that since the whites disbelieved the rumors about Pygmies in the neighborhood his friend had been loath to be laughed at or thought to be mad.

Professor Ledoux prevailed upon the eyewitness to come and see him and tell him his story.

> It had happened near the Meteorological set when they were taking their daily readings at 8 o'clock in the morning. Among the roots of a silk-cotton tree there suddenly appeared a little man with long reddish fur and long hair on his head—"same like white man"—but also reddish.

At once the little red man and the large black one took to their heels in opposite directions. For, according to the legends, the little forest-men brought bad luck. You only saw them once in a lifetime and you had to be alone.

I went to the place with my two informants. It lay in the shadow of thick forest, but was not too overgrown since the silk-cotton tree grew near a path. It was very likely that if there had been anything there it would have been easy to see.

It seemed unthinkable that within 12 miles of a big town like Abidjan, and 500 yards from huts inhabited by six Europeans and some 300 Africans, there could be unknown creatures in forest which though thick was far from virgin. All the same, Professor Ledoux decided to make discreet inquiries about the native legends but he did not obtain much information:

For, while there were plenty of men who "had seen" (?) them, they were reticent on the subject, always concluding that they were probably mistaken for all the encounters had taken place at nightfall. This is likely enough.

There was one relatively exact fact. In March 1946 a team of workmen . . . together with a European of whom I can find no trace, were supposed to have seen one of these little red men, at about 8 in the morning, in a tall tree in a very wooded little valley about half a mile from the future site of the station. The European asked what it was and the Negroes explained what a rare thing it was to see such a creature and the evil effects of doing so.

He was also told that during one of his expeditions in the course of 1947 the great elephant hunter Dunckel killed a peculiar primate unknown to him; it was small with reddish-brown hair and was shot in the great forest between the Sassandra and Cavally rivers.

In 1951 the professor's new boy told him that when he was young, probably around 1941, he had himself seen a hunter at Seguéla bring back a little man with red hair in a cage. The local official had put clothes on it for decency's sake and sent it to Abidjan by way of Bouaké. The boy did not know what happened to the little prisoner afterward.

Professor Ledoux remarks on the very firmly held belief that there are Pygmies in the forest between the Sassandra and Cavally rivers.

According to an African technician of mine . . . there was recently a system of barter between the Negroes and these forest creatures; various manufactured goods were left in the forest in exchange for various fruits. This was supposed to have gone on until 1935. The little men who practised this barter were hardly known even to the Negroes themselves. The Guérés called them *Séhité*.

The professor also found that the Europeans who had never spent any length of time between the Sassandra and the Cavally denied out of hand that there could be any little men in the forest, while those who had lived in this area were seriously prepared to consider that Pygmies might have lived there in the past and also that there might be a real basis for the legend of the red dwarfs. His own impression was that the legends and rumors in the Ivory Coast were based on the fairly recent presence of Pygmies and the present existence of reddish-haired primates whose exact nature was still problematical.

Although these "little hairy men" agree with nothing known to the zoologist or the anthropologist, their description could not agree better with a creature well known to paleontologists by the name of Australopithecus, which was still living in South Africa not more than 500,000 years ago, at a time when all the existing African species were already formed.

71. Reconstruction of the Australopithecus by Roy Chapman Andrews (1945).

The Australopithecus was a little creature, about 4 feet high, which walked upright on its hind legs and was proportioned like a miniature man. The first skull was discovered in 1924 embedded in a block of stone in a cave in British Bechuanaland and was sent to Professor Raymond Dart, who gave its owner the name of *Australopithecus transvaalensis*. He maintained that this creature had a much better claim than the Java ape man to be considered as the "Missing Link" between the great apes and man.

At first sight the skull seems to be merely a juvenile chimpanzee's, and a study of the teeth shows that this specimen was about six years old. But a more thorough examination reveals that the cranial capacity of 500 cc. is much greater than that of a chimpanzee. Moreover, the hole in the bottom of the skull through which the spinal cord passes is farther forward than in the anthropoids. This shows that the head, like a man's, sat plumb on the top of the spinal column. The pelvis also confirms that this little large-brained chimpanzee walked on its hind legs. Other evidence also seemed to show that its intelligence was high. In 1948 Professor Dart found fairly complete remains of an Australopithecus and with these fossils were some remains of charred bones and some charcoal. The professor therefore concluded that the Australopithecus knew how to use fire, on the strength of which he named it *Australopithecus prometheus.*

There was no sign it had stone tools, but Professor Dart thought that it made clubs of bones. Baboons' skulls, smashed in a peculiar way, and found with the Australopithecus's remains, seemed to imply that it ate its victims' brains. Australopithecus skulls were also found to be similarly crushed, which led some to suppose that this little Prometheus was also a Cain and a cannibal. Other authors suggested that the baboons and the Australopitheci had *both* been killed and eaten by man—much to the indignation of the South African anthropologists, who are determined that their little protégé should be the "Missing Link," the direct ancestor of man. However, today many paleontologists admit that the Australopithecus must have been contemporary with man,* indeed man has been blamed for its extinction. It occupied the same environmental niche as man, and man destroyed it. This is supposed to have happened no more than 500,000 years ago.

But if the Australopitheci were really so intelligent, would they have let themselves be killed without flinching? Surely, like many less cunning animals, they would try to move out of range of an obviously stronger enemy by taking refuge in a less agreeable but safer area.

In the thick forests of Africa, the last refuge of so many "relic" species, might not the Australopithecus's descendants still survive? Certainly the *agogwe* are extraordinarily like Dart's "intelligent, energetic, erect, and delicately proportioned little people." And

* This was confirmed in 1954 by a study of the sites by Dr. Kenneth P. Oakley, an anthropologist from the British Museum, who believes that the Australopithecus did not live in the caves, but was killed and taken there by beasts of prey. The layer of ashes—what is solemnly called the "basal hearth"—upon which the name *prometheus* rests, was analyzed and shown to be bats' dung which had probably ignited spontaneously!

the account of them mixing freely with baboons agrees with the fact that the two creatures' remains are so often found together. Perhaps there might be some sort of "gentleman's agreement" between these two very distantly related primates to live together in a sort of symbiosis, so frequent among animals.

If the *agogwe* are an unknown race of human Pygmies living with baboons, they could give us most valuable information about the most primitive of societies and strangest of partnerships. If they prove to be surviving Australopitheci, our present notions of man's origin would surely be radically changed. We should at last know exactly what these so-called ape men really were. These are important problems which deserve to be solved.

CHAPTER **17**

THE DRAGON ST. GEORGE DID NOT KILL

David Doldrum dreamt he drove a dreadful dragon. If David Doldrum dreamt he drove a dreadful dragon, where's the dreadful dragon David Doldrum dreamt he drove?

Traditional tongue twister.

AT HALF-PAST ELEVEN on the morning of December 23, 1919, Captain Leicester Stevens boarded the Southampton train at Waterloo station on his way to hunt the Brontosaurus in Central Africa.

His only companion was his faithful Laddie, a former "barrage dog," and he put his trust in his Mannlicher rifle, for the huge reptile was supposed to have a vital spot.

"Where that spot is," remarked the new St. George with an unduly mysterious air as the train moved off, "is one of my secrets."

This story is not a science-fiction writer's dream but is taken from the press of the time. To go dinosaur hunting in the twentieth century is surely madness—yet plenty of people were infected by it. For several weeks the most exciting news in the papers were Carpentier and Beckett warming up for their famous boxing match and that two different travelers had recently met a prehistoric monster in the Belgian Congo.

On November 17, *The Times* had published a small news item:

A TALE FROM AFRICA

Semper aliquid novi

The *Central News* Port Elizabeth correspondent sends the following:

The head of the local museum here has received information from a M. Lepage, who was in charge of railway construction in the Belgian Congo, of an exciting adventure last month. While Lepage was hunting one day in October he came upon an extraordinary monster, which charged at him. Lepage fired but was forced to flee, with the

> monster in chase. The animal before long gave up the chase and Lepage was able to examine it through his binoculars. The animal, he says, was about 24 ft. in length with a long pointed snout adorned with tusks like horns and a short horn above the nostrils. The front feet were like those of a horse and the hind hoofs were cloven. There was a scaly hump on the monster's shoulder. . . .

The dinosaur which the animal described in this dispatch most resembles is the Triceratops, which was some 19 to 26 feet long, and did have a horn on its nose. But it had "horns like tusks" rather than the other way round, and all four feet were thick and heavy like a pachyderm's. For some reason (mere journalistic ignorance is the most likely one) everyone began calling it a Brontosaurus, though this dinosaur has a long neck and tail, and neither horns nor tusks.

The Brontosaurus seems to have first reared its head in this affair in a delayed Reuter message from Bulawayo, dated December 4 but not printed in the London papers until the twelfth.

> News apparently corroborating the report of the existence in the Congo of a monster known as a *Brontosaurus* (the thundering Saurian) comes from Elisabethville.
>
> A Belgian prospector and big game hunter named M. Gapelle, who has returned here from the interior of the Congo, states that he followed up a strange spoor for 12 miles and at length sighted a beast certainly of the rhinoceros order with large scales reaching far down its body. The animal, he says, has a very thick kangaroo-like tail, a horn on its snout, and a hump on its back. M. Gapelle fired some shots at the beast, which threw up its head and disappeared into a swamp.
>
> The American Smithsonian expedition was in search of the monster referred to above when it met with a serious railway accident, in which several persons were killed.

Now the Brontosaurus, which is a reptile, is certainly not "of the rhinoceros order." And the animal described is no more like a Brontosaurus than a hedgehog is like a horse. But the news that so important a scientific body as the Smithsonian Institution had sent an expedition on the beast's track would perhaps have allayed the skeptics' doubts. And there was no doubt that this institution had in fact sent an expedition to explore this little-known corner of Africa, and that three of its members had been killed in a railway accident.

Soon the press was buzzing with the tale that as this disaster had brought the expedition to an end so prematurely, the Smithsonian

Institution had offered a reward of a million pounds to anyone who brought the mysterious monster back dead or alive. No more encouragement was needed for men like Captain Stevens to take up Brontosaurus hunting. Walter Winans, a big-game hunter of the highest reputation (according to the press), said in an interview that the famous animal merchant, the late Carl Hagenbeck, had told him before the war that two of his travelers on different expeditions at different times had seen a Brontosaurus in the swamps of Central Africa. These travelers were used to recognizing animals at a glance and their evidence ought to be trustworthy. Winans also drew a picture which was published in the *Daily Mail* with a caption which must have made the paleontologists of the time burst their sides with laughing or their blood vessels with just indignation:

> Mr. Winans, famous as a big game hunter and a world authority on animals, says that the Brontosaurus is a reptile, which crawls on its belly. He believes it is built very light in front, with short rudimentary legs.

Meanwhile the press was adding embellishments to the story almost daily. Lepage and Gapelle were now said to have been together when they saw a beast "like a lizard or a crocodile" which reared on its hind legs when it saw them. Captain Stevens "declared he saw it crashing through the reeds of a swamp, and that it was the Brontosaurus—a huge marsh animal, ten times as big as the biggest elephant." The number of members of the Smithsonian expedition killed in the railway accident increased dramatically, and this institution was said to have raised a fund of $5,000,000 to finance the monster's capture.

Finally Wentworth D. Gray wrote to *The Times*:

> Sir, I am authorised to contradict the statement that the members of the Smithsonian African Expedition who proceeded to this territory came here to hunt the brontosaurus. There is no foundation for this statement. I may also state that the report of the brontosaurus arose from a piece of practical joking in the first instance, and, as regards the prospector "Gapelle," this gentleman does not exist except in the imagination of a second practical joker, who ingeniously coined the name from that of Mr. L. Le Page.

This was enough to discourage the most earnest of dinosaur hunters and it was twelve years before a dinosaur hunt found its way into the press again.

In 1932 a young South African big-game hunter called F. Grobler returned to Cape Town after five years in Central Africa and told the press about a monstrous lizard which lived in the huge marshes where the Belgian Congo, Angola, and Northern Rhodesia meet:

> It is known by the native name of "Chepekwe." The natives in Central Africa used to call it the water lion.
>
> It can best be described as a huge leguan [iguana], the weight of which is estimated at about four tons or more. It was discovered about six months ago by a German scientist in the Dilolo swamps in Angola, and while I was in that country I saw photographs of it.
>
> I went to Lake Dilolo myself to look for it, but I did not see it. The natives say it is extremely rare and seldom seen, but they are convinced as I am of its existence.
>
> It lives only in swamps, and from what I was told it attacks rhino, hippo and elephant. I have seen a photograph of the "Chepekwe" on top of a hippo it had killed.

Grobler also told the press that he had acted as guide to Hans Schomburgk, who had gone to the Dilolo marshes in the wild hope of taking some sensational film shots of the *chepekwe*, but they did not manage to see the beast.

Grobler's good faith seemed indisputable, and in fact Schomburgk had said during a lecture in Germany the previous February that there was a native tradition that a huge reptile existed in Central Africa. Then a Swede, J. C. Johanson, wrote to the press:

> On February 16 last I went on a shooting trip, accompanied by my gun-bearer. I only had a Winchester for small game, not expecting anything big. At 2 p.m. I reached the Kassai valley.
>
> No game was in sight. As we were going down to the water, the boy suddenly called out "elephants." It appeared that two giant bulls were almost hidden by the jungle. About 50 yards away from them I saw something incredible—a monster, about 16 yards in length, with a lizard's head and tail. I closed my eyes and reopened them. There could be no doubt about it, the animal was still there. My boy cowered in the grass whimpering.
>
> I was shaken by the hunting-fever. My teeth rattled with fear. Three times I snapped; only one attempt came out well. Suddenly the monster vanished, with a remarkably rapid movement. It took me some time to recover. Alongside me the boy prayed and cried. I lifted him up, pushed him along and made him follow me home. On the way we had to traverse a big swamp. Progress was slow, for my limbs were still half-paralysed with fear. There in the swamp,

the huge lizard appeared once more, tearing lumps from a dead rhino. It was covered in ooze. I was only about 25 yards away.

It was simply terrifying. The boy had taken French leave, carrying the rifle with him. At first I was careful not to stir, then I thought of my camera. I could plainly hear the crunching of rhino bones in the lizard's mouth. Just as I clicked, it jumped into deep water.

The experience was too much for my nervous system. Completely exhausted, I sank down behind the bush that had given me shelter. Blackness reigned before my eyes. The animal's phenomenally rapid motion was the most awe-inspiring thing I had ever seen.

I must have looked like one demented, when at last I regained camp. Metcalfe, who is boss there, said I approached him, waving the camera about in a silly way and emitting unintelligible sounds. I dare say I did. For eight days I lay in a fever, unconscious nearly all the time.

Dinosaur hunters had become better journalists since the days of Lepage and Gapelle, not only in making the most of a sensational story but also in realizing the value of pictures. Alas, the picture published in several newspapers and magazines is a crude fake. A Komodo dragon has been transplanted somewhat clumsily into an African swamp, where it is perched on the tips of its toes on the carcass of a hippopotamus or rhinoceros, thus making it look enormous. Once again the dinosaur has proved to be a hoax.

All the same, the hoaxers were inspired by rumors founded on authentic facts. Hagenbeck really held the beliefs he was credited with, and Schomburgk had certainly been the first to talk to Grobler about the *chepekwe*, a beast which he had first heard of at the beginning of the century and had mentioned in a book in 1910. Grobler had done no more than to confuse or embroider the facts and make the mistake of thinking that Johanson's letter and photograph were a confirmation of them.

I shall do my best to explain what these facts are—though it will not be easy. We must take care not to mix up reports from widely separated areas nor to assume that all the tales of an "African dragon" necessarily refer to the same animal.

Let us begin a tour of the Belgian Congo, arranging the evidence by place rather than time. We shall start with Alfred Aloysius Horn, who has this to say about a large water monster:

Aye, and behind the Cameroons there's things living we know nothing about. I could 'a made books about many things. The *Jago-Nini* they say is still in the swamps and rivers. Giant diver it means. Comes out of the water and devours people. Old men'll tell you what their grandfathers saw, but they still believe it's there. Same as the *Amali* I've always taken it to be. I've seen the *Amali*'s footprint. About the size of a good frying-pan in circumference and three claws instead o' five. . . .

In 1902 Sir Harry Johnston, the man to whom we owe the discovery of the okapi, drew attention in his great work on Uganda to the mysterious creature in Lake Victoria, the third largest lake in the world.

There are also persistent stories amongst the natives that the waters of the Victoria Nyanza are inhabited by a monster (known to the Baganda as "Lukwata"). This creature, from the native accounts, might either be a small cetacean or a large form of manatee, or, more probably, a gigantic fish. So far, however, only one European has caught a glimpse of this creature.

This was Sir Clement Hill. C. W. Hobley writes:

The late Sir Clement Hill was proceeding from Kisumu to Entebbe in a steam launch some years ago, when off Homa Mountain a beast appeared out of the water and tried to seize a native sitting on the bow of the launch. It did not succeed, and Sir Clement (who was under the awning aft) told me that he distinctly saw it some little distance behind the boat; only its head visible, it was of a roundish shape and dark in colour. He was quite certain it was not a crocodile.

Hobley had heard natives on both sides of the lake talk of the *lukwata.* At first he thought these tales referred to big pythons sometimes found in the local folklore, but later realized it must be an unknown animal.

Edgar Beecher Bronson tells how he was hunting in the country west of Sotik when he met another hunter named J. A. Jordan, who told him of a monster called the *dingonek,* 14 or 15 feet long, with a head the size of a lioness's, but marked like a leopard's, with two long, straight, white fangs in the upper jaw. It was covered with scales like an armadillo, as broad as a hippopotamus, and spotted like a leopard. It had a long and broad tail. Its footprints were as big as a hippo's, but had claws like a reptile's.

At the time this story appeared [observes C. W. Hobley] it was considered that this was probably a traveller's tale, told to entertain a newcomer, but I have since met a man who a few years back was wandering about the Mara River . . . He emphatically asserts that he saw this beast . . . floating down the river on a big log, and he estimated its length at about sixteen feet but would not be certain of the length as its tail was in the water. He describes it as spotted like a leopard, covered with scales, and having a head like an otter; he did not see the long fangs described by Mr. Jordan. He fired at it and hit it; it slid off the log into the water and was not seen again.

I made inquiries of the District Commissioner, Kisii, Mr. Crampton, and he wrote recently and said he had visited the Amala River and made inquiries from the Masai in the neighbourhood, and they knew of the beast, which they called Ol-umaina, and described it as follows: About fifteen feet long, head like a dog, small ears marked somewhat after the fashion of a puff adder,* has claws, short legs, short neck, is said to lie in the sun on the sand by the river-side and to slip into the water when disturbed; when in the water only its head is visible.

This description agrees fairly well with the *dingonek* if not with the *lukwata.* In the 1930s Dr. E. G. Wayland, Director of the Geological Survey of Uganda, stated that he had been shown a piece of the *lukwata*'s bone. The natives of the Kavirondo country told him that they firmly believed in the *lukwata*; it fought with crocodiles and during the Homeric struggle lost pieces of its skin, which were much prized as amulets. They said that the monster's bellowing voice could be heard a long way off; indeed, Wayland said he had heard it himself. He could not explain these roars unless they were made by the *lukwata*—whatever this beast might be.

H. C. Jackson tells of an enormous snake, the *lau*, which the Nuer told him lived in the swampy area of the sources of the White Nile.

Some say that it has a short crest of hair on the back of its head not unlike that of a crowned crane: others that it has long hairs . . . In certain years, particularly during the rainy season, its belly is said to gurgle like the rumbling of an elephant. . . .

The Nuer state that it inhabits holes in the banks of rivers or swamps, and spends most of its time in swamps. . . .

The Addar swamps extend for some 1,800 square miles: those of the Bahr-el-Ghazal and its tributaries must cover an area of many thousands of square miles. Neither of these vast morasses has ever yet been explored, and many parts of them have never even been visited. . . .

There are so many stories of this creature from places as far apart as Bahr-el-Arab, Addar swamps, Bahr-el-Ghazal and Bahr-el-Zeraf,

* The puff adder has no external ears. Perhaps Hobley means the small horns on the horned viper, but the text is by no means clear.

> that it is difficult to dismiss as untrue the existence of some hitherto unknown serpent not unlike a gigantic python. . . .
>
> The common factor in the various accounts given of this creature is that it differs in colour from a python, in the shape of its jaws, in its greater length and girth, and there is no doubt that it is held in the greatest dread by the Nuer. They have a definite name for it, and there seems no particular reason, if this serpent were not distinct from the python, why the Nuer should fear a twelve-foot Lau more than he does a twenty-foot python.

Sergeant Stephens, an old hunter now in charge of a considerable stretch of the telegraph line on the east bank of the Nile, told J. G. Millais, the great British naturalist, more about the *lau*:

> Natives who are said to have seen a dead one describe it as forty to one hundred feet long, with a body as big as a donkey or a horse. The colour is said to be brown or dark yellow, and not green and black like the python. On its vicious-looking, snake-like head it has large tentacles, or thick, wiry hairs, with which it reaches out and seizes its victims. If a man sees a Lau first, the creature at once expires, but if the Lau is the first observer the human dies on the spot.
>
> . . . Abrahim Mohamed, in the employ of the company, saw a Lau killed near Raub, at a village called Bogga. The man I knew, and closely questioned. He always repeated the same description of the monstrous reptile. . . .
>
> A short time ago I met a Belgian Administrator at Rejaf. He had just come up from the Congo, and said he was convinced of the existence of the Lau, as he had seen one of these great serpents in a swamp and fired at it several times, but his bullets had no effect. He also stated that the monster made a huge trail in the swamp as it passed into deeper water.

I do not see why the *lau* of the Upper Nile marshes and the *lukwata* of Lake Victoria should not be the same animal—or at least animals of the same genus, for the great East African lakes are connected by the Bahr-el-Djebel with the huge marshes around the sources of the Nile.

It also seems as if the *lau*'s evil reputation, if not its actual area of distribution, stretches to that part of the Belgian Congo nearest to the Upper Nile and Uganda, where its description is known, as will be seen from the following story told to my father by a close friend, Colonel Alex Godart, a Belgian hero of World War I.

At the beginning of 1912, when he was a lieutenant in the Belgian Congo Police, Godart was going upriver on the Ituri in order to set up an outpost on the highest navigable point on the river. He had pitched camp in the middle of the virgin forest, when he was surprised to feel a violent earthquake. He was even more surprised when the one soldier who formed his escort rushed off to the riverside with his rifle in his hand.

Some time later the Negro came back rather sheepishly and said as if to excuse himself:

"It had already gone back into the water."

"What? What do you mean?" asked Godart.

"The beast [*nyama*] which makes the earth shake when it comes out of the water—it has to be killed, but I didn't see it."

The lieutenant was amused at this charmingly naive explanation, but he did not want to humiliate the Negro, who had shown nothing but courage, so he asked him several questions.

"Have you ever seen one?"

"I haven't, but my brothers have."

"Oh, and what is it like?"

"It's a very big beast like a hippopotamus, but with a little head with feathers—and a crest like a cock's comb."

The reader will have recognized the "short crest of hair . . . not unlike that of a crowned crane," in the Nuer description of the *lau*. But Godart found this detail so absurd that he had to cut short his interrogation so as not to burst out laughing. All the same, he admired the Negro's courage in dashing off with only one cartridge in his rifle to attack what he thought was a terrifying monster.

Naturally this story does not prove that in the Ituri itself there is a beast related to the *lau* of the Upper Nile and perhaps to the *lukwata* of Lake Victoria. It may just be that echoes of their existence have reached this area. At all events the tradition refers to a fairly heavy animal—and one with feet—if it is to make the earth shake—as elephants, rhinoceroses, and hippopotamuses do—and be blamed for seismic earth tremors.

But none of these monsters are at all like dinosaurs. Their various descriptions might variously apply to a stumpy and heavy snake, a giant monitor lizard, an unknown species of crocodile, or even some very odd kind of fish.

It is only the rumors that have come from other parts of Africa over the last fifty years which have gradually suggested the notion that dinosaurs, or rather some dinosaurs with characteristically long necks, may survive.

In 1909, Lieutenant Paul Grätz crossed Africa from the Indian Ocean to the mouth of the Congo in a motorboat and visited Lake Bangweulu, where he noted:

> The crocodile is found only in very isolated specimens in Lake Bangweulu, except in the mouths of the large rivers at the north; but in the swamp lives the *nsanga,* much feared by the natives, a degenerate saurian which one might well confuse with the crocodile, were it not that its skin has no scales and its toes are armed with claws. I did not succeed in shooting a *nsanga,* but on the island of Mbawala, I came by some strips of its skin.

Hans Schomburgk was surprised to find that Lake Bangweulu, although an ideal habitat for hippopotamuses, had no trace of one on its banks. When he expressed his astonishment to his native guides, they told him that there was an animal in the lake which killed and ate the hippopotamuses, although it was smaller than they were. It was an amphibious creature, but apparently it never came ashore, for its footprints were never seen.

When he reached the Dilolo marshes, 500 miles farther west, Schomburgk heard similar tales about a beast called the *chimpekwe,* which he later told to Carl Hagenbeck, who duly wrote:

> Some years ago I received reports from two quite distinct sources of the existence of an immense and wholly unknown animal, said to inhabit the interior of Rhodesia. Almost identical stories reached me, firstly, through one of my own travellers, and, secondly, through an English gentleman, who had been shooting big-game in Central Africa. The reports were thus quite independent of each other. The natives, it seemed, had told both my informants that in the depths of the great swamps there dwelt a huge monster, half elephant, half dragon. This, however, is not the only evidence for the existence of the animal. It is now several decades ago since Menges, who is, of course, perfectly reliable, heard a precisely similar story from the negroes; and still more remarkable, on the walls of certain caverns in Central Africa there are to be found actual drawings of this strange creature. From what I have heard of the animal, it seems to me that it can only be some kind of dinosaur, seemingly akin to the brontosaurus. As the stories come from so many different sources, and all tend to substantiate each other, I am almost convinced that some such reptile must be still in existence.

But the great Carl Hagenbeck died on April 14, 1913, without succeeding in solving the mystery which had puzzled him for so long.

Fortunately a British settler called J. E. Hughes who had spent eighteen years on the shores of Lake Bangweulu published a book in 1933 in which he gives us more information about a beast which he calls *chipekwe*—a remarkably similar name to Schomburgk's *chimpekwe* although Lake Bangweulu is 500 miles from the Dilolo marshes.

> It is described as having a smooth dark body, without bristles, and armed with a single smooth white horn fixed like the horn of a rhinoceros, but composed of smooth white ivory, very highly polished. It is a pity they did not keep it, as I would have given them anything they liked for it.

Hughes also heard from H. Croad, a retired magistrate in Northern Rhodesia, some evidence which seemed to confirm the native reports. One evening Croad had pitched his tent beside a small but very deep lake. In the middle of the night he heard a loud splashing noise, and in the morning he found huge footprints of an utterly unknown kind on the bank.

Hughes moreover reports that Robert Young, after whom Lake Young was named, shot at something in this lake: "It dived and went away, leaving a wake like a screw steamer." Young thought that the *chipekwe* still survived there; Hughes himself was more skeptical, and came to the conclusion that the *chipekwe* survived until very recently—only a few years ago if Croad's evidence was to be believed—but that it was now extinct.

Millais also writes:

> Mr. Denis Lyell . . . is convinced that there is, or was till recently, some large pachyderm, somewhat similar in habits to the hippopotamus, but possessing a horn on the head, which frequents the great marshes and lakes of Benguelo, Mweru, and Tanganyika. He calls it a water rhinoceros, and can adduce good evidence for his theory.

I think Lyell and Hughes are being too timid, for why should unknown animals suddenly die out as soon as there is good reason for thinking they exist? Clearly Lyell's "water rhinoceros" is the same as the *chipekwe*, since it agrees in its habits, size, and single horn. It also agrees in its area of distribution.

Here we must call the evidence of C. G. James, who wrote to the *Daily Mail* during the pseudo-Brontosaurus affair.

> Sir, I should like to record a common native belief in the existence of a creature supposed to inhabit huge swamps on the borders of the Katanga district of the Belgian Congo . . .

> It is named the *chipekwe;* it is of enormous size; it kills hippopotami (there is no evidence to show it eats them, rather the contrary); it inhabits the deep swamps; its spoor (trail) is similar to a hippo's in shape; it is armed with one huge tusk of ivory.

Professor Bonnivair tells me how he learned from Maurice Vermeersch that around the Lualaba, which links lakes Mweru and Tanganyika, the Baluba believe in a beast they call *nzefu-loi,* or "water elephant," not that it is necessarily a true elephant of semiaquatic habits. It is an amphibious animal with a body almost as big as a hippopotamus's, but with a very long neck and a hairy tail like a horse's. It has short and heavy ivory tusks—probably like a walrus's—and this may be why it is called "water elephant."

From the beast's description Bonnivair could not help suggesting that it might be a belated dinosaur of the Brontosaurus type, though the horse's tail should have raised some doubts. Indeed, examined objectively, none of the trustworthy reports that I have mentiond so far justifies us in using the word "dinosaur." In many cases it is by no means certain that the beast is even a reptile: it is more like a mammal. Hagenbeck talks of "a huge monster, half elephant, half dragon," using the word "dragon" in the same sense as the Komodo dragon. Unscientific reasons seem to have caused the elephant half to be deliberately forgotten.

John G. Millais produced one of the few significant reports about an African animal which might be a sort of dinosaur. It comes from the remains of the Barotse "Empire," on the middle Zambezi.

> The late King Lewanika, who was much interested in the study of the animals of his kingdom, Barotsiland, frequently heard from his people of some great aquatic reptile, possessing a body larger than that of an elephant, and which lived in the great swamps near his town. He therefore gave strict orders that the next time one was seen he should be told, and he would at once go himself and visit the place. In the following year three men rushed into his court house one day in a great state of excitement, and said they had just seen the monster lying on the edge of the marsh, and that on viewing them it had retreated on its belly, and slid into the deep water. The beast was said to be of colossal size, with legs like a gigantic lizard, and possessing a long neck. It was also said to be taller than a man, and had a head like a snake.
>
> Lewanika at once rode to the spot and saw a large space where the reeds had been flattened down, and a broad path, with water flowing into the recently disturbed mud, made to the water's edge. He de-

scribed the channel made by the body of the supposed monster to Colonel Hardinge, the British Resident, "as large as a full-sized wagon from which the wheels had been removed."

The Boers' wagons were about 4 feet 6 inches wide. Millais does not say what the Barotse called this beast, but Captain William Hichens made inquiries and found that the natives always called it *isiququmadevu.*

Continuing clockwise around the Belgian Congo, we come to eastern Angola, by following up the Zambezi toward the Dilolo marshes, where the *chipekwe* lives. It is not surprising to find tales of amphibious monsters there.

Ilse von Nolde, who spent 10 years in eastern Angola, reported in 1930 that the natives believed that there was an amphibious monster known in Kimbundu as *coje ya menia* or "water lion." This may be because of the sound of its roar. They often drew her attention to distant rumbling sounds they said were made by this beast, which generally lived in the water but quite often climbed out onto the bank. Thus the Angola "water lion" roars like the *lukwata* and *lau* of the Nile marshes; it also slaughters all the hippopotamuses, even though it is smaller than they are, like the *chipekwe* of the sources of the Congo. And like the *nzefu-loi* of the Lualaba it is said to have tusks or large canine teeth which it uses for killing.

A Portuguese truck driver told Frau von Nolde that one day on the Cuango around 9° S. latitude he learned that a *coje ya menia* had chased and killed a hippopotamus the night before. He went off with several native hunters and found the tracks of this chase. It consisted of the tracks of a hippopotamus that was obviously in a hurry, mixed with unknown tracks smaller than a hippopotamus's, a little like an elephant's but within the circular print there was "the mark of toes beneath the sole of the foot." At last the trackers reached a place where the grass and bushes had been crushed and smashed, and in the middle of this battlefield lay the carcass of a hippopotamus hacked and ripped as if with a huge bush knife. Nothing had been eaten.

Crossing the Congo estuary we come full circle in the Cameroons.

In 1913 they were still a German colony, and Kaiser Wilhelm's Government sent out a reconnaissance expedition, the Likuala-

Kongo Expedition, to map the country and report on its mineral and vegetable wealth. In charge of the expedition was Freiherr von Stein zu Lausnitz, a captain in the colonial forces. His official report was never published, because Germany lost interest in the Cameroons when she lost the colony itself after World War I, but it still exists in manuscript, and Willy Ley has published a passage about the *mokéle-mbêmbe*. Captain von Stein points out that his information is based on the accounts of "experienced guides who repeated characteristic features of the story without knowing each other."

> The animal is said to be of a brownish-gray color with a smooth skin, its size approximating that of an elephant; at least that of a hippopotamus. It is said to have a long and very flexible neck and only one tooth but a very long one; *some say it is a horn.* A few spoke about a long muscular tail like that of an alligator. Canoes coming near it are said to be doomed; the animal is said to attack the vessels at once and to kill the crews but without eating the bodies. The creature is said to live in the caves that have been washed out by the river in the clay of its shores at sharp bends. It is said to climb the shore even at daytime in search of food; its diet is said to be entirely vegetable. This feature disagrees with a possible explanation as a myth. The preferred plant was shown to me; it is a kind of liana with large white blossoms, with a milky sap and apple-like fruits.

Ivan T. Sanderson writes to me that in 1932 he came upon "vast hippo-like tracks" on the Upper Cross River above Mamfe Pool, where there were no hippos because this creature—whose name he transcribes from Anyang as *mbulu-eM'bembe*—drove them away. Some months later,

> Gerald Russell and I with two boys in two small canoes entered this river from below through a gorge into the Mamfe Pool at dusk, as we passed some all-but-submerged caves in the cliffside, got the shock of our lives when the most terrific noises I have heard outside of warfare issued from one of the said caves and *something* (and it was the top of a head we both feel sure) much larger than a hippo rose out of the water for a moment, set up a large wave and then gurgled under.

In the northern Cameroons the natives also fear a water monster, which, though it seems very different from the *mokéle-mbêmbe*, is reminiscent in name at least of the *coje ya menia* or "water lion" of eastern Angola. This is told by another German colonial called Naumann of Ulm.

> The Kaja who live among the granite rocks were astonished to see me ride a horse. They had never seen one. This is what they told my interpreter:
>
> In their country there was also a big beast with a large mane which lived in the water. The animal was very dangerous, it was a much feared wild beast. It was only very rarely seen. Its name in the Baya language was *dilai,* or "water-lion."
>
> Although I promised a reward of 50 marks, no one was able to show me this animal. In another area they were called "water-leopards."

I have had independent confirmation of this information, and also had it much completed by Lucien Blancou, formerly Chief Game Inspector in French Equatorial Africa. "Throughout the Ubangi-Shari," he writes, "there are traditions and tales concerning at least two hitherto unknown animals." One is what the Baya call *dilali* (and not *dilai*)—which in fact does mean "water lion" and the Banda *mourou-ngou.* The other is called *diba* by the Baya and *badigui* by the Banda. It is said to be a sort of huge snake which occasionally strangles hippopotamuses but does not eat them.

Let us see what Blancou can tell us about the first of these mysterious water beasts. On May 26, 1930, he had shot his first hippopotamus in the heart of Ubangi-Shari, deep in the bush, several days' march from the nearest villages.

> The animal still had not floated at nightfall, so we camped nearby until the following day waiting for the carcass to come to the surface. During the night there was a high wind and small rain. At dawn the porters and trackers told me that they had heard a *mourou-ngou* calling near the dead hippo. When the beast had been dragged ashore I saw that there were signs that the carcass had been bitten, apparently by crocodiles. But all my men knew crocodiles well. Unfortunately they had not thought it necessary to wake me so that I could hear the "water panther's" cry.

In 1932 in eastern Ubangi-Shari an interpreter told Blancou that the *dilali*—though he had not seen it himself—had a horse's body and a lion's claws. A native guard of Zandé origin who was present during the conversation added that it had large tusks like a walrus's. He maintained that it ate fish, though the Baya said it ate nothing but leaves, and it was shown not to be a carnivore by an old chief's story that they had found a hippo and a crocodile, some two miles from the river, which had been bitten by the *dilali* but were otherwise intact. Blancou was also told that the *mourou-ngou* sometimes even killed elephants. An old man called Moussa told a story which Blancou relates as follows:

> In 1911 (this date has been cross-checked) when he was porter with a detachment of riflemen going from Fort Crampel to Ndélé, Moussa saw one of these soldiers seized by a *mourou-ngou* at the junction of the Bamingui and the Koukourou. The animal was shaped like a panther, a little larger than a lion but with stripes, and about 12 feet long. The background of its coat was likewise the color of a panther's, but its footprint was oddly described as containing a circle in the middle(?).
>
> The soldier was in a canoe when the animal came out of the Koukourou "like a hippo," just where the rivers met, seized the man in the canoe and dragged him into the water capsizing the boat, surfaced once more with the soldier in its mouth and then disappeared. The man paddling the canoe swam safely away, but the soldier's rifle and kit remained on the bottom of the river. . . .
>
> This story struck me at once, because I had just been checking the records of the outpost at Ndélé and found signs that a rifleman had been lost about this time.

In 1945, Blancou's gunbearer, Mitikata, drew the following sketch of the *mourou-ngou* for him. It was about 8 feet long. It had a small head, fangs like a lion's, a plump body which was brownish above and below, and a panther's tail. It lived in the water all the time, raising its head above the waves only in the evening, and killed men by dragging them into deep water.

In 1937 the village chief of Ouanda-Djalé, in eastern Ubangi-Shari, told Blancou about a mysterious beast called *gassingrâm* which haunted the district. He said it was larger than a lion and reddish brown all over. Moreover its footprints were larger. It had been seen abroad in the daytime, though very rarely, carrying off its prey to caves in the mountains. At night its eyes shone like lamps.

Blancou thinks this animal is the same as the *dilali* and the *mourou-ngou* because they are all described as large cats, but it seems to me that the striking difference in habitat confutes this view. To introduce a land animal into this chapter of amphibious monsters can only confuse it even more, and God knows it is confused enough already by animals very different from one another.

Let us proceed to the second unknown animal in the Ubangi-Shari, the snakelike *badigui,* also known as *ngakoula-ngou.*

It was described to Blancou by the Linda Banda as a gigantic snake that killed hippos without leaving a wound and browsed on branches of trees without leaving the water. This would of course be a very odd diet for a snake, which would not be able to lift its head out of the water unless it was resting on the bottom, so one may well suspect that there might be a bulkier body behind the snakelike head and neck. Like the *isiququmadevu* and the *mokéle-mbêmbe,* it really does remind one of a dinosaur.

In 1928, one of these animals was said to have crushed a field of manioc belonging to the chief and had left tracks between 3 and 5 feet wide. The same beast killed a hippopotamus at about the same time in the Brouchouchou River.

The Baya gave Blancou a similar description of the monster, and in 1934 he learned more precise and definite details about the beast from old Moussa, already quoted.

> When he was about 14 years old [Lucien Blancou relates], and the whites had not yet come (about 1890, I suppose), Moussa was out laying fish traps with his father in the Kibi stream . . . It was one o'clock in the afternoon in the middle of the rainy season. Suddenly Moussa saw the *badigui* eating the large leaves of a *roro,* a tree which grows in forest galleries. Its head was flat and a bit larger than a python's (Moussa spread his hands and put them together to show me the size). Its neck was as thick as a man's thigh and about 25 feet long,* much longer than a giraffe's; it had no hair, but was as smooth as a snake, with similar but larger markings. The underneath of its neck was lighter—also like a snake's. Moussa did not see its body. . . .
>
> According to him, the old men believed the *badigui* was dangerous, but he had never heard of anyone else meeting one. He had seen its tracks out of the water, but only on the bank, and they were as broad as a python's. Finally he said that the *badigui* does not frequent places where you find hippopotamuses, for it kills them.

* This figure is, of course, Blancou's, based on Moussa's indications; "much longer than a giraffe's" is a more valuable yardstick. Now a giraffe's neck is 7 or 8 feet long. If the *badigui*'s neck was really 25 or even 15 feet long Moussa would surely have said "two or three times as long as a giraffe's." I think a length of 10 or 12 feet is a reasonable interpretation of Moussa's evidence.

In 1945 Mitikata, whom I have already mentioned, told Blancou about the *ngakoula-ngou,* saying that it was a huge snake which left tracks as broad as a truck, and that he had seen one of these tracks 15 years before.

So much for the reports. Do they all refer to the same animal? Hardly. But they are all amphibious. Indeed, this has been my criterion in including them in this chapter. They are also not so very different in size. All the reports refer to an amphibious animal with a body roughly as thick as a hippopotamus's—rather less, according to most of them. This has led some authors into thinking all the animals are the same. Yet in other respects they are very different. The skin is sometimes said to be covered with scales, sometimes bare or smooth. In some the neck is short; in others it is long, even very long and flexible. The tail is at one time compared to a crocodile's, at another to a horse's or to a panther's. Finally it is surprising to find that some have tusks or fangs like a walrus's, while others have a little horn on the nose and one even has a crest and barbels. In short, the various descriptions of African "dragons" obviously refer to at least three quite different types of animal—so it is not surprising that a mixture of these creatures has produced an absurd and fantastic beast, which it is difficult to believe in, let alone identify.

Let us deal first with the animals which seem to have short necks: the *dingonek* (or *ol-umaina*) of Kenya, and Lieutenant Grätz's "degenerate saurian," the *nsanga* of Lake Bangweulu. The *mourou-ngou* with its spots and projecting fangs is probably similar to the *dingonek* and should therefore be included in this first group.

Now all these are rather heavily built quadrupeds, between 12 and 16 feet long, reminiscent in shape of a crocodile, giant lizard, or theoretically even an ultragigantic otter. The *dingonek* or *ol-umaina* has several of a crocodile's features. It is covered with conspicuous scales and has a broad, long tail. It is spotted like a leopard, as crocodiles not infrequently are. But only with a great deal of exaggeration can a crocodile's largest teeth, like canines, be compared to a walrus's tusks. Moreover, the *dingonek*'s head has been compared to an otter's, a dog's, and a lioness's, and this hardly agrees with the Nile crocodile's characteristically long skull and jaws. And what are we to make of the sort of ears it is alleged to have, which may be like the horns of a horned viper?

It is only among prehistoric reptiles that we find true tusks. Without going so far as to maintain that it must be one of them, we can safely say that the *dingonek* belongs to a very special genus. In the jungle rivers of West Africa there lives the little black crocodile (*Osteolaemus*), which differs from the other African crocodiles, all of which belong to the genus *Crocodylus,* in being darker, smaller (not much over 6 feet long), and especially in having a shorter, rounder head like a cayman's. Perhaps, unknown to us, there may be a giant species, slightly larger than the Nile crocodile, and doubtless on the road to extinction. If small colonies of such a species survived in Kenya and Ubangi-Shari, it might explain the legends of the *dingonek* (or *ol-umaina*), and the *mourou-ngou.*

The *nsanga* alone of this first group cannot possibly be classed among the crocodiles, from which it differs in its smooth skin and in having claws. A smooth-skinned crocodile with claws reminds one irresistibly of a giant monitor like the Komodo dragon. As the monitors have fairly long necks—at all events longer and more flexible than those of crocodiles—the *nsanga* will serve as a link with the next group of African "dragons," those which, in part at least, are shaped like snakes: the *lau* of the Upper Nile swamps, the *nzefu-loi* of southern Belgian Congo, the *isiququmadevu* of the Upper Zambezi, the *mokéle-mbêmbe* of the Cameroons, and the *badigui* of Ubangi-Shari.

We can set the *lau* quite apart from the rest of this group, from all of which it differs in so many peculiar features that its identity seems to me quite transparent. It is described as a thick, short water snake, with a crest of "hair" and barbels on its head; it makes a loud, rumbling noise, comes out of the water and leaves a broad track on land, and lives in holes burrowed in the banks.

Apart from the beast's size, this description perfectly fits a fish of the catfish family (Siluridae), of which there are many species in all the muddy waters of tropical Africa. The catfishes may be recognized by the long barbels around the very wide mouth. Some have a dorsal fin which might be mistaken for a crest. They all tend to have long bodies, and some look exactly like eels—or snakes. Others can breathe air on the surface, and come out of the water and crawl on land in search of animal food. *Clarias lazera* of the Sudan even spends the dry season in burrows, from which it also emerges on the prowl at night. And *Eutropius* has long air sacs as further aid to respiration, and makes a growling noise which can be heard a long way off. Finally, certain catfishes have poisonous spines which can cause appalling wounds, and there is one, the electric catfish called "thunderfish" by the Arabs, which produces violent electric shocks. It would be hardly surprising if such fish were much feared, and it was eventually believed that you could be thunderstruck if some of them so much as looked at you.

There remains the *lau*'s alleged size. In Africa none of the known catfishes is very large: the electric catfish is the biggest

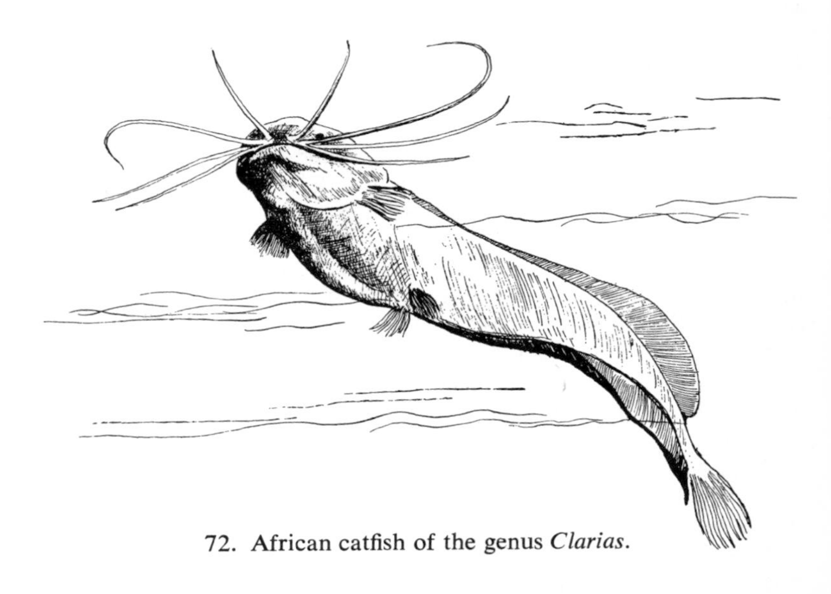

72. African catfish of the genus *Clarias*.

at 5 feet long. But the giant catfish of European rivers is sometimes more than 10 feet long and weighs more than 400 pounds. Professor Osmond P. Breland writes with his usual caution: "This fish is said to attain a length of fifteen feet and a weight of over 700 pounds, but I have not been able to confirm these figures." It seems to be the largest known fresh-water fish. It is enormously voracious, attacking not only other fish, but also small land animals like birds and mammals. Thus it would not be so strange if there were an unknown and perhaps formidable giant catfish in Africa.

The *lukwata* of Lake Victoria, which Sir Harry Johnston thought was "more probably a gigantic fish," is also very likely a huge catfish. Sir Clement Hill, who saw one, said that it had a rounded, fishlike head. The head of a catfish 4 feet long may be as much as 16 inches wide, so it is easy to see how terrifying the gaping maw of a similar fish would be if it were two or three times the size.

The *nzefu-loi* or "water elephant" is a particularily thorny question, but it may give a clue to part of the problem. Since it is alleged to have a hairy tail like a horse's, it cannot very well be a reptile. But why on earth *should* it be one? The reference to a "very long neck" has suggested the notion of some sort of Brontosaurus. May it not really be an elephant, as its name implies, one of those little "water elephants" that I have already discussed? The "water elephant's" neck is longer than that of the ordinary elephants, and it certainly has hair like horsehair on its tail, as all the elephants do. Moreover, may not the tales of amphibious monsters as much like an elephant as a hippopotamus or "half elephant, half dragon" be founded on the same kind of animal?

Here it may help if I quote the first circumstantial account of elephants of aquatic habits, which one Le Petit gave a British naturalist called R. J. Cuninghame. During the five years he spent traveling in the French and Belgian Congo, Le Petit twice saw water elephants. The first time was in 1907, and they were swimming in the Congo River. He saw them a second time in the swampy area between Lake Tumba and Lake Leopold II.

> On the second occasion [Cuninghame reports], he saw five of these animals on land and was able to look at them for fully one minute through his glasses and, as they were in tall grass 400 yards distant, he had a good opportunity to observe. He took a shot at one of them and hit it in the shoulder, but though he offered the natives an ample reward they never were able to recover it.
>
> Height at shoulder, 6 to 8 feet. Legs relatively short. Back curved as in *E. africanus*. Tail not observed. Neck *about* twice the length of *E. africanus* with ears similar in shape to those of that species, but relatively smaller. Head most distinctly long and ovoid in form, with trunk only about 2 feet in length. The shape of the feet was seen in the spoor on sand and showed four toes distinctly separated as in the hippopotamus, but the weight of the body seemed to be carried by the toes largely, while the plantar impression of the sole was not very pronounced. The ground was level where this spoor was seen. All the animals observed had no traces of tusks. Skin is apparently hairless, smooth and shiny resembling that of a hippopotamus, only darker. The gait was elephantine, and the last seen of those five water-elephants was their disappearance into the water, which was deep.
>
> In habit they are nocturnal, coming out to feed on strong rank grass after sundown. They spend the day in the water much as hippopotami do. The Babumas (fishing natives) know this animal well and have a name for it, "Ndgoko na Maiji," meaning the water-

elephant, and they fear it greatly as it is known to rise from the water and with its short trunk capsize canoes. It is also very destructive to the nets and reed fishtraps of the natives. Its locality is apparently very restricted, and the natives maintain they are not very numerous.

Cuninghame goes on to remark that the primitive forms of the Proboscidea had a "long-shaped, ovoid type of skull" like modern tapirs and that "the cervical vertebrae were antero-posteriorly much longer than is now the case in modern elephants." He concludes: "These two characteristics are very marked in the water-elephants, and M. Le Petit is most emphatic that the heads reminded him more of enormous tapirs than of any other existing animal."

Le Petit's strange Proboscidea are remarkably like certain primitive and supposedly extinct elephants, which nevertheless

73. Reconstruction of the Dinotherium.

survived until very recently in Africa, where they were certainly contemporary with man. The huge Dinotherium, with tusks curving downward like a walrus's, was one of them. The largest species (*D. giganteum*) was 16 feet high at the shoulder. The remains of a smaller species (*D. hopwoodi*) have been excavated from Pleistocene deposits not more than 100,000 years old in Kenya and Tanganyika.

At the end of the last century some zoologists refused to admit that the Dinotherium was quite extinct. Might not a small aquatic species still survive? Frau von Nolde says of the usual description of the *coje ya menia* or "water lion" of the Upper Cuango that it "always seems to apply to some pachyderm armed with powerful teeth which it uses as weapons," and the animal's tracks were described to her as like an elephant's with "the mark of toes beneath the sole of the foot." And the *dilali* or "water lion" had tusks like a walrus's.

All these "water lions"—a name which may be an allusion to their ferocity, power, or sovereignty over other beasts—are supposed to kill hippopotamuses, but not to eat them. It would not be at all surprising to find a water elephant behaving in this way. The hippopotamuses would occupy the same habitat and claim the same territory and food, so they would be particular enemies. It is easy to see how the water elephant might chase a hippopotamus ashore, stabbing savagely with its tusks, and eventually leave it in the state in which the beast was found that had been so appallingly lacerated by a *coje ya menia.* This is certainly the way a Dinotherium would fight, with its tusks pointing downward.

Now that we have eliminated a short-nosed crocodile, a giant catfish, and water elephant (which none the less deserve attention from the zoologists), we still have the large amphibious animals with very long necks. They alone are relevant to any discussion of the possible survival of a dinosaur in Africa.

First let us consider the *nsanga* and *chipekwe,* which can be included in the long-necked group only on the slenderest evidence.

The *nsanga*, as I have already mentioned, might be a large monitor lizard. Now, although the monitors may be compared to crocodiles in general shape, they have a relatively long and slender neck, especially in certain species. This and the apparently smooth skin are strictly speaking the only features in which the *nsanga*, if it is a monitor, resembles the *mokéle-mbêmbe* and the rest.

The *chipekwe* also seems to have several points in common with them. Not only is its skin smooth, but it has the odd reputation of killing hippos but not eating them. Only Schomburgk reports that it devours them, but this may be his own opinion, since all the other legends and reports unanimously contradict it. The *chipekwe*

74. Australian monitor lizard, showing its conspicuously long neck.

and *mokéle-mbêmbe* are, moreover, linked by a secondary anatomical feature: the single horn which they both have on the nose.

Yet if the *chipekwe* really belongs with the *mokéle-mbêmbe, isiququmadevu,* and *badigui,* it is odd that no one has mentioned that it has a long neck. Actually we know so little about this animal's appearance that it is impossible to say what it may be. Denis Lyell may have had good reasons for calling it a "water rhinoceros," so may Schomburgk for thinking it was "a sort of saurian," but neither of them has told us what they were.

So let us concentrate on the *mokéle-mbêmbe, badigui,* and *isiququmadevu.* The *badigui* is the least known of the three, but it may well be identical with the *mokéle-mbêmbe*, since they both have a snake's head, a long neck, and smooth skin. Moreover they both eat leaves, and though they attack and kill anything which annoys them, whether it be hippopotamuses or men in canoes, they do not eat them. They also live in neighboring areas, the Ubangi-Shari and the Cameroons, both watered by tributaries of the Ubangi.

The *mokéle-mbêmbe* and *isiququmadevu* are no less alike, though one lives in the Cameroons and the other in Northern Rhodesia, at least 1,200 miles apart. They both have a body almost the size of an elephant's, a long neck and long powerful tail like a crocodile's or a lizard's. The *isiququmadevu* has a snake's head, and both it and the *badigui* leave a track about 5 feet wide.

The *mokéle-mbêmbe*'s skin is said to be brownish gray. The *badigui*'s, on the neck at least, is marked with pattern like a python's but bigger. The only adornment on the three is the *mokéle-mbêmbe*'s single tooth, which is presumably a horn. It may well be an inconspicuous bump or a feature peculiar to the male, and might thus have been missed by some of the witnesses.

If we assume as a working hypothesis that the three amphibious animals are of the same kind, this gives us a picture of a large quadruped with a small, rather triangular head adorned with a little inconspicuous horn on the end, a long and flexible neck, a thick thorax a little bigger than a hippopotamus's, legs on the sides of its body, and a long and powerful tail. Its skin is smooth, brownish gray, and marked with a large dark pattern. From the bulk of its body let us say that it is roughly 25 to 28 feet long or even more than 30 feet if its tail tapers very gradually.

It is amphibious and spends most of its time in the water, making its lair in a hole in the bank. It also ventures on land in search of its food, which is basically vegetable. Although it is not carnivorous, its territorial instinct impels it to kill its rivals, the hippopotamus and probably the manatee too. No doubt the same instinct impels it to capsize canoes stupidly mistaking them for rivals too.

What, then, is its zoological identity? The first step is to assume that it is a reptile. But it need not necessarily be one. Denis Lyell thought the *chipekwe* was a sort of aquatic rhinoceros and he may be right in this case, but a long-necked monster, even if it has a horn on its nose like the *mokéle-mbêmbe*, can obviously not be a rhinoceros. So, assuming it is a reptile, what kind is it? In outline it is certainly like a dinosaur; that one can hardly deny. But it could be a giant species of monitor. These reptiles are found in all the hot countries in the Old World. The Australian monitor has a relatively long and slender neck; its cousin, the Komodo dragon, stands upon very substantial legs; and until very recently there was a giant monitor (*Varanus priscus*) 25 feet long in Australia.

All we can really say is that there seems to be a large and unknown reptile in Central Africa. Its identity remains to be determined, and in view of its size it is probably, like the crocodiles, the Komodo dragon, and the giant tortoise, a relic of the great reptile empire that flourished in the Jurassic period. I have already mentioned the remarkable geological and climatic stability in Africa, which could hardly be better suited for preserving such an ancient type.

The most surprising aspect of the Congo "dragon" has been brilliantly expounded by Willy Ley. He shows that pictures of the beast are found not only in Central Africa but even on one of the most remarkable of ancient Babylonian monuments, the Ishtar Gate.

In 1902, after three years of excavation, when Professor Robert Koldewey brought to light the splendid arch raised by King Nebuchadnezzar, he found that walls were decorated with glazed brick bas-reliefs representing alternate bulls and dragons. The bulls have been shown to be the aurochs (*Bos primigenius*), the terrible wild bulls which survived in Europe until the Middle Ages. The *sirrush*, as the "dragon" is called, is much more of a problem. It is a scaly quadruped with the front legs of a lion and the back legs of an eagle, with a snake's head on the end of a long neck. A forked tongue darts from its mouth, and it bears on its head a single horn, various wattlelike ornaments, and even a short mane.

Bizarre though this animal is, it seems to be intended to be a real beast, since it has been alternated with very realistic bulls, and in the passage leading to the gate there is a double row of no less realistic lions. The three beasts seem to have been chosen as the three most powerful and savage creatures known. What can the dragon be?

In his sumptuous work *Das Ischtar-Tor in Babylon*, published in 1918, Professor Koldewey was quite definite:

> If a creature like the *sirrush* existed in nature it would belong to the order of Dinosauria and the suborder of Ornithopods. The Iguanodon of the Cretaceous layers of Belgium is the closest relative of the dragon of Babylon.*

The Chaldeans were not paleontologists. The *sirrush* cannot be a reconstruction of a long-dead animal. Nor is it likely to be one of the many Babylonian mythical beasts, constructed of a hotchpotch of other beasts, for while they had a very short life in Chaldean art, the *sirrush*, like the lion and the aurochs, was depicted over a period of nearly 2,000 years from about 2800 B.C. to the reign of Nebuchadnezzar (1146–1123 B.C.).

There is also explicit reference to a live dragon at Babylon in the book of Bel and the Dragon in the Apocrypha:

> And in the same place there was a great dragon which they of Babylon worshipped.
>
> And the king said unto Daniel, wilt thou also say that this is of

* The *sirrush* is more like the Ceratosaurus in having a little horn, but I shall show later why this identification is untenable.

75. Reconstruction of the Iguanodon, a running dinosaur that was herbivorous.

> brass? lo, he liveth, he eateth and drinketh. Thou canst not say that he is no living god: therefore worship him.
>
> Then said Daniel unto the king . . . give me leave and I shall slay this dragon without sword or staff. The king said I give thee leave. Then Daniel took pitch, and fat, and hair, and did seethe them together, and made lumps thereof: this he put in the dragon's mouth, and so the dragon burst in sunder . . .

Whether or not this living "dragon" in Babylon is anything more than apocryphal, it does not seem to have served as a direct model for the one on the Ishtar Gate and on many other Babylonian monuments; for the *sirrush* is certainly not an accurate picture of the Iguanodon or Ceratosaurus. But it is a good stylized likeness.

The *sirrush* could have lived in Central Africa, where it has been proved that the Chaldeans went, and where they could have seen a giant lizard. When Hans Schomburgk came back to Europe with native tales about the *chipekwe*, he also brought several glazed bricks of the same type as those on the Ishtar Gate. Naturally he had no idea that the beast and the bricks could have any connection —which Willy Ley was the first to recognize. It is therefore possible that one of the amphibious species of Congo "dragon" was known to the Chaldeans 5,000 years ago from travelers' tales, and that this beast was the origin of the *sirrush* on the Ishtar Gate.

This can be demonstrated by a simple experiment. Ask someone who has never seen a picture of a dinosaur—a child, for example— to make a drawing from your own verbal description of the Congo "dragon." They will begin by drawing a quadruped, similar in outline to some well-known beast (horse or cow, dog or cat), and then go on to give it the long tail and snake's neck and head. When you mention that it has large claws on its feet they will give it a lion's or a cat's paws. They will add the required ornament to its head, and when you say that it is a reptile, they will cover it with scales. There is every chance that the result will look very like a *sirrush* apart from the hind legs. Repeat the experiment with another naive and amenable artist, this time drawing a picture of a dinosaur, such as Charles Knight's Ceratosaurus, Plate 24, following page 146, from a dictated description, and the result will probably be a *sirrush* again, even down to the birdlike hind feet.

All this seems to me to imply: (1) that the *sirrush* is based not on a direct knowledge of the Congo "dragon," but on a description derived from travelers' tales; (2) that the Congo "dragon," like the *sirrush,* had birdlike hind feet; (3) that both animals are a reptile very similar to some of the bird-footed dinosaurs. This does not prove that the Congo "dragon" is a dinosaur, but the theory should not be rejected out of hand. We may even make a guess as to what dinosaur it is.

The Ornithopods, the group to which the Iguanodon belongs, are not the only ones with birds' feet. The Theropods also have them, and include the Ceratosaurus, which has a little horn on the end of its nose, and which is thus particularly like the *sirrush*. But while the Ornithopods are herbivorous, the Theropods are formidable carnivores. The Congo "dragon" is a vegetarian, and cannot therefore be one of the Theropods. It is more like the Ornithopods, but not the Iguanodon, not only because it has a horn—a minor detail —but because it is essentially aquatic, while the Iguanodon is a running beast that normally stands on its hind legs.

Actually we only know a minute fraction of the dinosaurs that once inhabited the whole earth, and the Congo "dragon" does not exactly agree with a reconstruction of any that have been disinterred. If the argument based on the *sirrush* has any value, the beast must be an Ornithopod. If it has none, the animal is much more like a Brontosaurus, the animal with which by some strange irony of fate ignorant journalists confused the imaginary Ceratopsian invented by those hoaxers Lepage and Gapelle: it would not be a true Brontosaurus, of course, but a long-necked and much smaller Sauropod of the same family.

All this supposes that it *is* a dinosaur, which is very far from being proved.

PART SEVEN

THE LESSON OF THE MALAGASY GHOSTS

In one moment I've seen what has hitherto been
  Enveloped in absolute mystery,
And without extra charge I will give you at large
  A Lesson in Natural History.
    LEWIS CARROLL, *The Hunting of the Snark.*

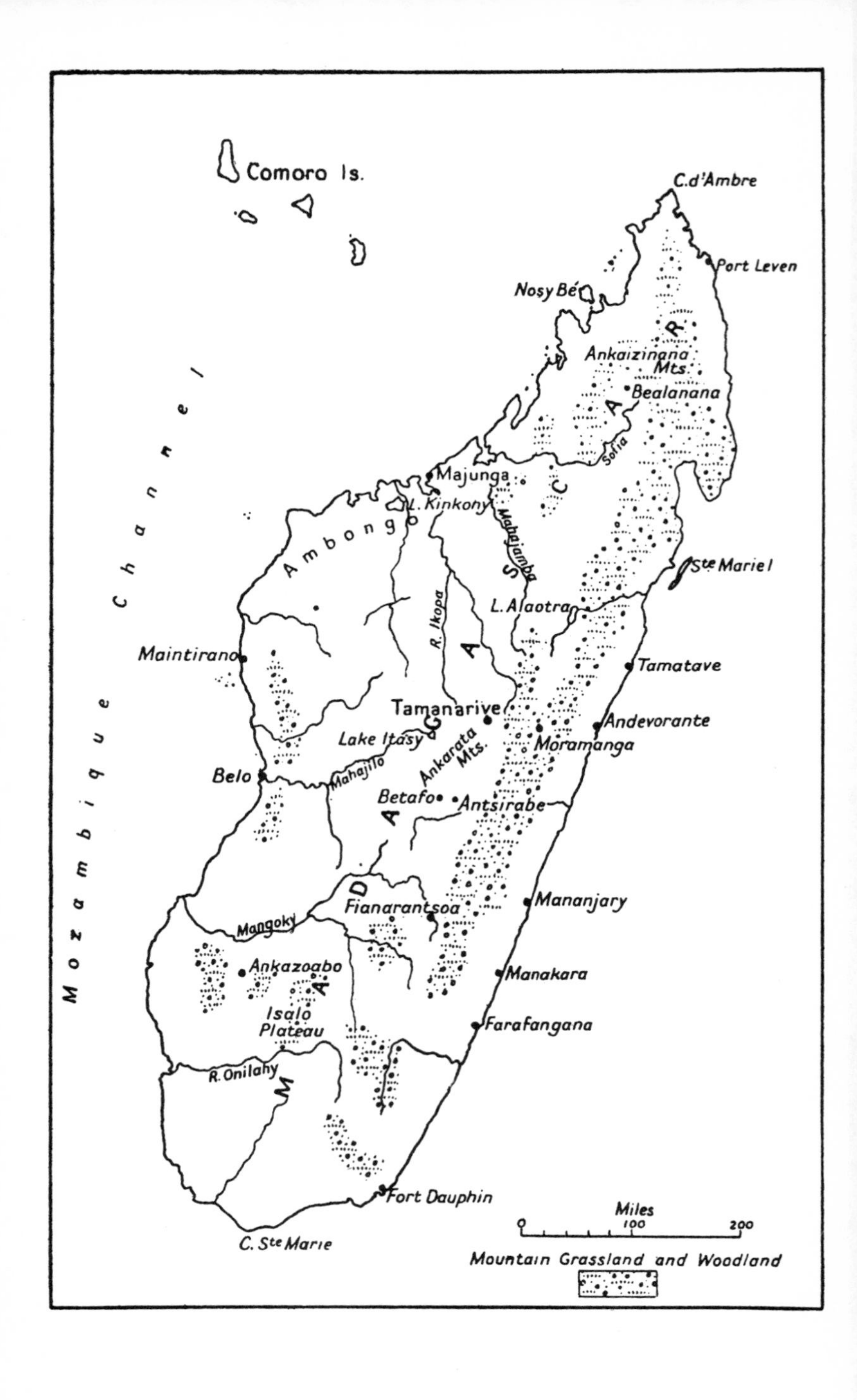

Comoro Is.
C.d'Ambre
Port Leven
Nosy Bé
Ankaizinana Mts.
Bealanana
Sofia
Majunga
L. Kinkony
Mahajamba
Ambongo
Ste Marie I
L. Alaotra
R. Ikopa
Maintirano
Tamatave
Tamanarive
Andevorante
Lake Itasy
Moramanga
Ankarata Mts.
Mahajilo
Belo
Betafo
Antsirabe
Mananjary
Fianarantsoa
Mangoky
Ankazoabo
Manakara
Isalo Plateau
Farafangana
R. Onilahy
Fort Dauphin
C. Ste Marie
Mozambique Channel
MADAGASCAR
Miles
0
100
200
Mountain Grassland and Woodland

CHAPTER **18**

TRATRATRATRA, VOROMPATRA, ET CETERA

Madagascar is another land of promise. Here, too, mountain and forest prevail; situation is favourable; and we know almost nothing of the interior.

P. H. GOSSE, *The Romance of Natural History*.

ONCE UPON A TIME there was a famous doctor who cured all his patients (this is a fable, of course); his one fault was that he was very incredulous. One day a little girl came to him, and begged him to hurry to her mother's sickbed.

"Your mother is all right, my child," he replied with a smile, "she imagines she is ill, but actually she's as sound as a bell. Don't worry, she will get well by herself."

"But, doctor, she is in terrible pain!"

"Never mind. Run along now, I have other things to do."

The little girl went. The mother died. And the doctor proved that it was not enough to have complete mastery of his art. His first duty was to examine his patients.

Many zoologists have this doctor's fault. While they remain obstinately deaf to rumors about unknown animals, these animals die out. Some are already extinct, and we have lost the opportunity of studying them alive. This thesis is amply borne out by the history of our knowledge of the animals of Madagascar.

When Admiral Étienne de Flacourt published his *Histoire de la Grande Isle de Madagascar* in 1658 after a long stay in that country, he mentioned:

> *Vouroupatra,* a large bird which haunts the Ampatres and lays eggs like the ostrich's; so that the people of these places may not take it, it seeks the most lonely places.

Subsequent travelers told fantastic tales about the bird's size, saying that its eggs were enormous, big enough for the natives to use as tanks for drinking water. Needless to say, hardly anyone believed these tales. The ostrich was already a monstrously large bird at 8 feet high. A bird that laid eggs eight times as large would have to be impossibly huge.

There had also been much earlier tales. Herodotus was told by Egyptian priests about a race of gigantic birds "beyond the sources of the Nile" which were strong enough to carry off a man. As the most powerful eagle cannot lift more than a rabbit or a small lamb, this would be a staggering feat. In the thirteenth century Marco Polo claimed that Kublai Khan had shown him a bird's feathers some 60 feet long and two eggs of prodigious size.* He said that he thought the roc came from some islands to the south of Madagascar. But the roc and the huge bird of Madagascar were assumed to be equally fabulous.

Nevertheless in 1832 the French naturalist Victor Sganzin actually saw an enormous half-eggshell in Madagascar. The natives were using it as a bowl but they would not sell it to him. In the meantime a traveler called Goudot found remains of similar eggs in Madagascar and showed them to Professor Paul Gervais, of the Paris Museum, who attributed them at first to a sort of ostrich.

In October 1848 John Joliffe, the surgeon of H.M.S. *Geyser*, made friends with a French merchant called Dumarele. Joliffe wrote:

> M. Dumarele casually mentioned that . . . at Port Leven, on the north-west of the island [Madagascar], the shell of an enormous egg, the production of an unknown bird inhabiting the wilds of the country, which held the almost incredible quantity of *thirteen wine quart bottles of fluid*!!!, he having himself carefully measured the quantity. It was of the colour and appearance of an ostrich egg, and the substance of the shell was about the thickness of a Spanish dollar, and very hard in texture. It was brought on board by the natives (the race of "Sakalavas") to be filled with rum, having a tolerably large hole at one end, through which the contents of the egg had been extracted, and which served as the mouth of the vessel. M. Dumarele offered to purchase the egg from the natives, but they declined selling it, stating that it belonged to their chief, and that they could not dispose of it without his permission. The natives said the egg was found in the jungle, and observed that such eggs were *very very rarely* met with, and that the bird which produces them is still more rarely seen.

Nobody took much notice of these reports; they were not much publicized and seemed too fantastic to be true. But in 1851 the existence of a giant bird in Madagascar was at last officially ad-

* The "feathers" may well have been palm leaves.

mitted when a merchant captain called Abadie found three eggs and some fragmentary bones on the southwest coast of the island. The two largest eggs were 13⅝ inches by 8½ inches and 12⅝ inches by 15⅜ inches, one being more elongated than the other. Their capacity was nearly 2 gallons. Each egg would have held six ostrich's eggs or 148 hen's eggs—enough to make an omelet for 70 people.

On the strength of these eggs and bones and with the precedent of Owen's *Dinornis,* described in 1839, Isidore Geoffroy-Saint-Hilaire christened it *Aepyornis maximus,* or "the tallest of the high birds." This name was not entirely deserved. The tallest of the birds was not the Aepyornis but one of the moa tribe, which reached a height of over 11 feet. Geoffroy-Saint-Hilaire deduced that the Aepyornis must be 16 feet high, partly because its egg was double the dimensions of an ostrich's. Actually the Aepyornis very rarely reaches 10 feet: it is a bulky rather than a tall bird.

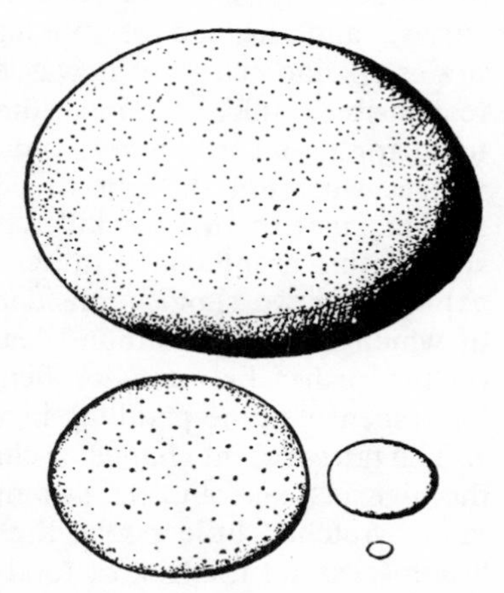

76. Aepyornis's egg compared with an ostrich's, hen's, and hummingbird's.

In 1866 Alfred Grandidier fished out of pools at Ambolisatra some huge bones in a perfect state of preservation. At first sight they seemed to belong to some large pachyderm, but when examined they proved to belong to the mysterious Aepyornis which was at once nicknamed "elephant bird." Soon enough other bones were disinterred for the huge bird's whole skeleton to be deduced. The complete skeleton of *Aepyornis maximus* reconstructed at the Paris Museum is 8 feet 9 inches high—a very respectable size for this bird. There is no constant ratio between the size of a bird and the eggs it lays. The kiwi's eggs are as big as a medium-sized ostrich's, 5 inches by 2¾ inches, although the bird itself is hardly any larger than a hen. Its body is only three times as heavy as its egg. All the same, even if *Aepyornis maximus* is not the tallest known bird, it is the largest and weightiest. It has been deduced that the *Aepyornis maximus* weighed 965 pounds, while a 10-foot moa weighed only 520 pounds.

There is little doubt that it was the bird that De Flacourt described by the name of *vouroupatra*. Does this mean that it was still alive then? The eggs found around the great pools in the south and southwest were sometimes so fresh that they seemed to have been laid quite recently. The bones were equally fresh. And the natives said that the giant birds lived in the thickest forests of the island but were very rarely seen.

Unfortunately most naturalists were still so soaked in Cuvier's ideas that they refused to believe that such a monstrous bird could survive and, instead of looking for the bird itself, racked their brains to find out why it was extinct. Man could not be directly responsible. Unlike the moa, the *vouroupatra* had not been slaughtered for its meat. There is no local legend to this effect, and the natives were not thought to be well-enough armed to hunt it—though, since the Maoris killed moas with nothing but stone-headed spears, this argument is rather weak. But to find the most likely explanation we should consider the bird's habits. The conditions in which its eggs are found lead one to believe that it laid them on the rushes like a moor hen. From its anatomy it might well have spent its time paddling in wooded swamps.

The fairly recent change of climate dried up the immense lakes in the high plateaus and the swampy areas. Forced to huddle in ever more wretched little pools, the Aepyornis would have eventually become extinct for lack of food and suitable shelter. The heaps of their remains in peat bogs support this theory.

No doubt the thoughtless deforestation perpetrated by man had a decisive effect on the change of climate. The consequences of the gradual destruction of the virgin forest would have taken quite a long time to make themselves felt, and until 1867, if not later, the natives maintained that the *vouroupatra* still lived in some remote areas. Today there is no hope of finding a living specimen, because the swampy forest has decreased so dramatically.

If the naturalists in the last century had not been so incredulous of Malagasy legends and travelers' tales, but had searched systematically for De Flacourt's *vouroupatra*, we might know much more about this extraordinary bird which may have aped the hippopotamus in its habits, as it did in its structure.

The change in climate was also said to be the death knell for several giants of very different groups; so that today Madagascar's fauna consists only of small or medium-sized species. But is this necessarily true? The change in climate was not sudden and was so recent there is still a faint hope of finding some survivors of these giants of the past.

Madagascar is the land of subfossil creatures. The excavated remains of huge animals, whether they be birds, reptiles, or mammals, are sometimes so fresh that one cannot help wondering whether the species they come from are really extinct. The native traditions often make it perfectly clear that they were known by man until quite recently, for, according to Van Gennep's principle, which I have already quoted, the memory of an event dies out in two centuries when there are no written traditions. This makes the old accounts of voyages so valuable to our knowledge of recently extinct animals.

Admiral Étienne de Flacourt did not only speak of the Aepyornis as a living bird. He also mentioned another animal which no zoologist has yet been able to trace:

> *Trétrétrétré* or *Tratratratra* is an animal as big as a two-year-old calf, with a round head and a man's face; the forefeet are like an ape's, and so are the hind feet. It has frizzy hair, a short tail, and ears like a man's. . . . One has been seen near the Lipomani lagoon in the neighborhood of which it lives. It is a very solitary animal, the people of the country are very frightened of it and run from it as it does from them.

Most naturalists, of course, took this animal to be mythical, for there are no true monkeys in Madagascar and therefore no man-faced beasts with ape's feet. De Flacourt's tale was no more to be trusted than his repetition of the fable about dog-headed men that had been told by Ctesias, Marco Polo, and many others. The skeptics received their first blow when the indris (*Indris brevicaudatus*) was found to exist. It is the largest of the lemurs known today and is extraordinarily like a little man with a dog's head. Three feet high and with no tail but an inconspicuous stump, the indris is astonishingly like a man in outline. Like the other lemurs, or half-monkeys, it has a fine and pointed muzzle, which makes its head more like a fox's or a dog's.

Since the legend of the dog-headed men proved to be so well founded, more notice should have been taken of De Flacourt's report of the *tratratratra*. I don't know whether anyone ever looked for it "near the Lipomani lagoon"; certainly they never found it.

Paleontological discoveries were, however, to prove that it might very well have existed, but first I must say something about the way Madagascar's fauna reached that country through the ages and how it evolved.

One might reasonably expect to find the same fauna in Madagascar as in Africa except for a few insular differences. Actually the animals on the island are sometimes quite original and sometimes related to creatures from as far afield as South America and

77. The indris, the largest known living lemur.

southeast Asia. The most typical of Madagascar's fauna are the lemurs, which, like the monkeys, are primates, but have many distinctive features in their anatomy. The lemurs are so closely connected with Madagascar that few people realize that at the beginning of the Tertiary era they were found all over the world. They are often found in Eocene strata in North America and even in Europe. From Madagascar's size and position one might think that it had been the last refuge of these defenseless beasts. But the lemurs were not driven to Madagascar; it was the home from which they spread over the world, where they were almost all eventually ousted by the monkeys. Just then Madagascar was cut off from Africa and the island which had been their cradle remained their undisputed empire. Although three quarters of the species of lemurs are found in Madagascar, there are galagos, pottos, and angwantibos in Africa, and the slender and slow loris in Asia, which have managed to survive by all becoming nocturnal. In Madagascar itself there are almost as many species that are active by day or at dusk as there are that are thoroughly nocturnal.

Safe from the large beasts of prey, almost without herbivores, and innocent of any monkeys, Madagascar was a perfect place in which the lemurs could expand and flourish and become almost as varied as the marsupials in Australia.

This phenomenon of differentiation is not very striking in the species that are still extant. The subfossil lemurs, which survived until quite recently, were much more varied in their habits and structure. The Megaladapis, with its concave rhinoceroslike skull and vast jawbones, was a true tree-dwelling pachyderm. Although it was as big as a cow it must have been able to haul itself slowly through the thickest of trees with its long, strong, and prehensile fingers. The Paleopithecus evolved by convergence an ape in Madagascar very similar to the African chimpanzee. The Hadropithecus aped man himself. Professor C. Lamberton writes:

> Its round head, its broad flat face, its almost straight nose standing well clear of the face, its eyes close together, the upright symphysis of its lower jaw all tend to make it look not merely ape-like, but almost like a human.

The Hadropithecus had stout limbs and probably usually moved on the ground. The bones of other, no less extraordinary lemurs have been found, and there may well be many more still undiscovered. De Flacourt's description of the *tratratratra* no longer seems fantastic, and one must admit that it could well apply to what we know of the Paleopithecus, for example.

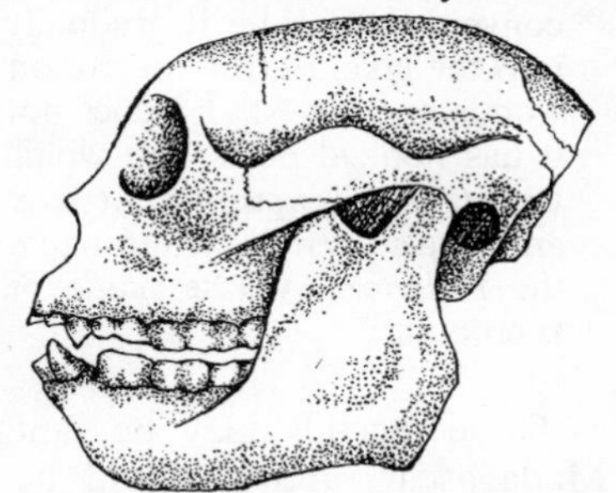

78. Skull of Hadropithecus with its astonishingly human profile.

What of the other Malagasy tales of animals unknown to science? What, for instance, is the *tokandia* in the old saying "the *tsomgomby* goes straight ahead but the *tokandia* moves in jumps"? Legend says that it is a huge jumping quadruped, which climbs into the trees, where it lives. Its face is not like a man's, but it cries like a man.

Surely it is not pure chance that this barely credible legend of a large arboreal quadruped should be found in the only country where such an animal—the Megaladapis—has ever lived?

And what are we to make of the *kalanoro* that Raymond Decary speaks of?

> All the tribes believe in the existence of little creatures similar to our gnomes and elves. Their name varies with the district: *Biby olona, Kokolampo, Kotokely,* and so on. The *Kalanoro* is, if I may say so, a sort of amphibious variety. In Lake Aloatra, the *Kalanoro*

are women like naiads or mermaids, with hair that falls to their waists; they live at the bottom of the water in watertight palaces, seducing canoers and fishermen and luring away children. The Betsileo, on the other hand, believe that *Kalanoro* is a little land dwarf, which is not more than two feet high, entirely covered with long hair and is the wife of another dwarf called *Kotokely; Kalanoro* lives in caves where she lies on a fairy's bed made of silkworm cocoons. She is interested in the children of men, stealing them and substituting her own young; she makes the babies she has stolen drink special potions which prevent them from growing. Ill-favoured new-born children are sometimes called "sons of *Kalanoro,*" and in the country the parents still keep careful watch at a birth to see that *Kalanoro* does not succeed in making an exchange.

Around Lake Kinkony the Sakalava have a very different notion of the *Kalanoro*. This rather more masculine creature lives in the thickets and reeds on the edges of lagoons. It is less than three feet high; it has long hair and only three toes on its feet. It has a sweet woman's voice, lives on fish and raw food, and walks in the country in the evening. If you meet one, it will accost you and hold you in conversation while it gradually leads you away until you disappear into the lake. Farther north, on the other hand, the *Kalanoro* lives in woods and caves, and does not try to lead human beings astray, but it has hooked nails with which it can give cruel wounds to anyone who tries to capture it; it lives on milk, which it sometimes comes and steals even from the natives' huts. In short, young and old fear the *Kalanoro,* whose name parents invoke to make their children keep quiet.

These legends may be fantastic, but they are found all over Madagascar, and it would be odd if they were utterly without foundation. Professor Lamberton wisely says:

> The freshness of the bones found in the south-east, as well as the state of the deposits, seems to show that the Hadropithecus must have lived until quite recently. . . . According to the Bara, the *Kalanoro* are little long-haired men who still live in the forests in the district and come out at night to prowl the villages in search of food. They run and climb very nimbly.
>
> All this agrees well enough with what we might expect of the Hadropithecus, and it is not impossible that the natives who lived in this area only a few centuries ago saw these animals, the memory of which has been handed down from generation to generation, gradually being altered to become more and more mysterious.

But as there are still some 8,000,000 or 9,000,000 acres of virgin forest, it would be rash to assert categorically that not a single giant lemur survives. After all, the okapi, which is as big as a horse, remained unknown for a long time because of its nocturnal habits.

Raymond Decary, who has written about the mysterious beasts of Malagasy legend at some length, thinks that most of them are mythical, though they derive in part from the subfossil hippopotamuses and lemurs "which haunted the forests and waters only a few centuries ago, and the memory of which has been confused, giving rise to legendary beliefs." But there are others which he thinks look more real—the *habéby*, for instance, which the Betsileo call *fotsiaondré*, or white sheep. It is a purely nocturnal beast, which is said to be sometimes seen by moonlight in the uninhabited wastes of the Isalo range. It is as large as a sheep and is also supposed to have a cloven hoof and a white coat, somewhat spotted with buff or black. It has a long muzzle, long furry ears, and large staring eyes.

Personally I very much doubt whether the *habéby* is really a sort of sheep, for sheep are not in the least nocturnal. Besides, no one seems to have ever seen a *habéby* ram, at all events they have never mentioned horns. On the other hand, some of the lemurs are curiously specialized, and the *habéby*'s large and staring eyes, which all witnesses insist upon, are not at all like an ungulate's, though they are a striking feature of the lemurs, especially the nocturnal species, which are the most ghostlike of all.

Admiral de Flacourt reports yet another puzzling Malagasy animal:

> *Mangarsahoc* is a very large beast, which has a round foot like a horse's and very long ears; when it comes down a mountain it can hardly see before it, because its ears hide its eyes; it makes a loud cry in the manner of an ass. I think it may be a wild ass.

Ten years later the Comte de Modave wrote:

> This animal is much feared by the Negroes, who give terrifying descriptions of it; at bottom it is but a wild ass; many are found in this part of the island, but you must look for them in the woods, for they never leave these lonely places and are hard to approach.

Decary says that in Anosy they still speak of an animal called *mangarisaoka* (literally "whose ears hide its chin"), which is of course the same as De Flacourt's *mangarsahoc*. A wild ass has several times been reported in the Ankaizinana forests and the Bealanana and Manirenjy districts. Tracks of its hoofs have been found. This harmless beast (if it really is an ass) has a vile reputation among the natives, who think that the mere sight of it will bring bad luck.

Several naturalists are almost ready to admit that in Madagascar there is a wild ass or horse which no white man has ever seen and of which not the slightest remains have been found, yet they flatly refuse to consider the possibility that any large lemurs can have survived. These nimble, intelligent creatures which live in forests and are nocturnal in habits would have a much better chance than a horse or ass of eluding our searches. And their ghostlike, owlish eyes would be able to arouse the greatest superstitious terror among the natives. The survival of subfossil animals in Madagascar is denied out of the same obstinate prejudice that I have attacked throughout this book. It all stems from an obsolete scientific doctrine, Cuvier's Revolutions of the Globe.

In Madagascar fortunate circumstances have almost enabled us to watch the extinction of the giant fauna of the past, but we have missed our opportunity. We should take the lesson to heart, for there are still numbers of strange or giant forms of animals hiding in the more remote corners of the world, as is shown by the persistent and often exact reports that I have recorded here. Are we to let these monsters die out without attempting to know them? Surely the inquiring spirit of Aristotle and Pliny, Gesner and Aldrovandus, Buffon and Darwin is not dead?

In a sense I hope that it may be. Now that I have reached the end of this book I feel a vague sense of regret—regret that I have revealed the still-undisturbed retreat of so many unknown animals. When I think of what man has done with a rifle, I am horrified that I should offer him new targets.

The most dreadful of the monsters I have mentioned attacks man only in self-defense or to provide himself with food. Only man kills for pleasure.

No sooner is a new animal discovered than the hunt for trophies begins. These maniacs must be stopped at once. History shows how alarmingly quickly man can exterminate a whole species. We have hardly known our noble cousin the gorilla for more than a century, yet, despite the tireless efforts of those institutions set up to protect him, he has been ignominiously slaughtered, and perhaps will soon vanish without our having learned anything about his habits, so full of significance from our own point of view.

Tomorrow we may know one of our other relatives: the abominable snowman, for instance, who is surely a shy and gentle great ape; or perhaps an even more human primate like the tiny *agogwe* or the elusive *orang pendek*. I hope with all my heart that when he is captured there will be no needless murder.

Have pity on them all, for it is we who are the real monsters.

Index